ReMake It!

Recycling Projects from the Stuff You Usually Scrap

BY TIFFANY THREA

D0117720

♪ STERLING

New York / London
www.sterlingpublishing.com/kids

For my craftabulous niece, Marina. —T.T.

STERLING and the distinctive Sterling logo are registered trademarks of Sterling Publishing Co., Inc.

Library of Congress Cataloging-in-Publication Data Available

Lot#:
2 4 6 8 10 9 7 5 3 1
11/10
Published by Sterling Publishing Co., Inc.
387 Park Avenue South, New York, NY 10016
© 2011 by Tiffany Threadgould
Illustrations © 2011 by Tiffany Threadgould
Distributed in Canada by Sterling Publishing
c/o Canadian Manda Group, 165 Dufferin Street
Toronto, Ontario, Canada M6K 3H6
Distributed in the United Kingdom by GMC Distribution Services
Castle Place, 166 High Street, Lewes, East Sussex, England BN7 1XU
Distributed in Australia by Capricorn Link (Australia) Pty. Ltd.
P.O. Box 704, Windsor, NSW 2756, Australia

Printed in China
All rights reserved.

Sterling ISBN 978-1-4027-7194-1   4538 7734 3/11

For information about custom editions, special sales, premium and corporate purchases, please contact Sterling Special Sales Department at 800-805-5489 or specialsales@sterlingpublishing.com.

This book is printed on FSC-certified paper.

VELCRO® is a registered trademark of Velcro Industries B. V.
etchall® is used with permission from B&B Etching Products, Inc.
etchall® is non-toxic and safe for children with adult supervision.

Packaging products featured in this book are used with permission from Amy's Kitchen, Inc., Barbara's Bakery, Happy Herbert's Food Co., Inc., Chelsea Milling Company, *Make Magazine* and O'Reilly Media, and Ronnybrook Farm Dairy.

Designed by Merideth Harte, Mina Chung, and Amy Wahlfield.
Photos by Kevin Schaefer. Author photo on page 128 by Abby Kelly.

Contents

Get Ready to
ReMake It!

Have you ever used an empty glass jar as a drinking glass or turned an old tin can into a pencil holder? If the answer is yes, then you've already had fun ReMaking It!

This book presents simple ideas for reusing stuff you were just about to throw away, showing you step-by-step instructions on how you can ReMake things all by yourself! Just add some of your own style or use the tips in this book to make the coolest one-of-a-kind items. The best part is that most of the materials you need are free and can be found at home. Some of these projects might require extra craft supplies from a store. But, for the most part, you'll be able to make these crafts using things you can find all around you.

Did you ever think the oval opening in the lid of a tissue box could become a picture perfect photo frame? Or how about the sleeve of an old T-shirt becoming a nifty little drawstring pouch? It's a cinch! After reading *ReMake It!* you'll never look at trash in the same way ever again.

When doing these projects, be sure to ask permission before ReMaking something that you or your family might still use. Don't cut up things like your favorite T-shirt, your dad's tie, or a record album case before asking a grown-up if it's something you can ReMake into new. Certain projects might require an adult's supervision, so please be sure to ask for help when needed.

And when you are done making all the crafts in this book, you can go to the website **www.sterlingpublishing.com/ReMakeIt** to find free downloadable bonus projects.

Let's start ReMaking!

Stitchionary:
Basic Sewing Terms & Techniques

FABRIC SCISSORS

Fabric scissors are scissors with sharp blades made for cutting fabric.

NOTCHING CORNERS AND CURVES

Having curves or corners without notches may make the fabric bunch up. This will make it hard to turn the project inside out and make it difficult to get smooth, finished edges.

There are two kinds of curves to your fabric: inner curves and outer curves. And there are two techniques for handling them: clipping and notching.

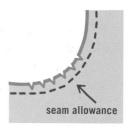

seam allowance

Clipping the fabric enables it to fold over itself and lie flat on the inside of the project. To clip a curved seam, use your scissors to make small cuts into both seam allowances around the curve. Be careful not to cut into the seam you've stitched.

seam allowance

seam allowance

Notching the seam allowances on an outer curve or corner enables fabric to stretch comfortably around the curve and allows the seam to lie flat. To make a notch, cut tiny triangles into the seam allowance. Be careful not to cut into the seam you've stitched.

RIGHT SIDE VS. WRONG SIDE

The right side of the fabric is what will be on the outside of your finished project. The wrong side is the other side—like the inside of a T-shirt or a pair of jeans. The wrong side often looks slightly faded and has the seams showing.

SEAM ALLOWANCE

The seam allowance is the amount of space between the raw edge of the fabric and your stitch. In this book, seam allowances are all ½ inch unless otherwise noted.

Stitchionary:
Stitches you'll use to ReMake It

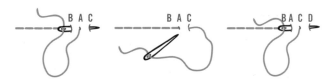

BACKSTITCH

BLANKET STITCH

Backstitching prevents your seams from unraveling and gives extra strength where it is needed.

To sew a backstitch, thread a needle and tie a double knot at the end of the thread. Bring the needle up through the fabric from the wrong side (stitch A). Insert the needle back through the fabric a short distance behind stitch A (this makes stitch B) and pull the thread through until the stitch is flat. From the wrong side again, bring the needle through the fabric a short distance in front of stitch A (this makes stitch C) and pull the thread through. Bring the needle back to just in front of stitch A on the right side of the fabric. Insert the needle and pull the thread through. From the wrong side, bring the needle up through the fabric in front of stitch C. Repeat these steps, keeping your thread flat against the fabric for a smooth stitch. On the wrong side of the fabric, the stitches will be overlapping in a straight line.

A blanket stitch is a decorative stitch used along the edge of a project—like along the edge of project **#66 Felted Sweater Potholder**.

Thread your needle with embroidery floss and tie a double knot. Start a stitch ¼ inch in from the edge of the fabric by inserting the needle from the back of the fabric to the front. Bring the needle and thread over the top of the fabric edge, back to the starting point of the first stitch, and through the same hole. Keep the loop made over the edge of the fabric loose. Put the needle through the loop made at the top of the fabric edge. From the back of the fabric, make another stitch about ¼ inch from the first one, from back to front. Now, pull the thread flat. Repeat these steps, always remembering to thread the needle through the loop of the loose stitch before pulling it flat to the fabric.

RUNNING STITCH

The running stitch is the most basic stitch. Sew straight or curved seams with small consistent stitches. Make the stitches the same size and distance from each other on both sides of the fabric.

To sew a running stitch, thread a needle and tie a double knot at the end of the thread. Bring the needle up through the fabric from the wrong side, then back through the fabric, then back up again. Make stitches that are consistently about ⅛ inch long and make the space between all stitches consistent.

TOPSTITCH

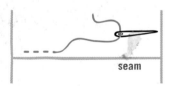

seam

Topstitching is used both to hold pieces firmly in place and to add a more finished look. To topstitch, make a simple line of stitching on the outside of a fabric, a small distance from the finished seam. A topstitch is just like a running stitch, except it always falls right next to a seam.

ZIGZAG STITCH

This stitch is decorative and a fun addition to your projects.

To sew a zigzag stitch, thread a needle and tie a double knot at the end of the thread. Bring the needle up through the fabric from the wrong side (stitch A), then back through the right side of the fabric at a 45-degree angle (stitch B). Make another stitch from the wrong side of the fabric right next to stitch B (stitch C). Stitch back through the right side of the fabric at another 45-degree angle (stitch D). Repeat these stitches at alternating 45-degree angles to create a zigzag pattern. Make stitches that are consistently about ⅛ inch long.

Chapter 1
Paper Pizzazz

1. Scrap Paper Picture Frame

MATERIALS:

- 2 sheets of scrap paper
- scissors
- ruler
- tape
- photo
- clear plastic from packaging

You can use different types of scrap paper for many of these paper projects. Try using magazine pages, comic books, last year's calendar, newspaper, sheet music, maps, and more. Look in your paper recycling bin for inspiration.

INSTRUCTIONS:

1. Measure and cut two pieces of paper into squares that are 8 inches on each side.

2. Take one square piece of paper and fold it in half corner to corner. The paper should now look like a triangle. Fold the triangle in half again, straight across the middle. Unfold the paper so it is flat again.

3. Take all four corners of the paper and fold them into the center. Open the piece of paper back up. This is fold A (see diagram a on the opposite page).

4. Now, fold each corner twice toward the center. First, fold each corner so the point touches fold A. Then fold again along fold A so the corner is hidden. There will now be a 3-inch x 3-inch open square in the center of the folded parts (see diagram b).

5. Flip your project over (keeping all the folds you've made in place). Fold each new corner back 1 inch toward the center (see diagram c). Set this piece aside.

6. With the other piece of paper, make the picture frame stand. Start by folding the paper in half long-ways down the center. It should now look like a long rectangle. Open the piece of paper so it's flat.

7. Fold each side of the paper lengthwise to the center fold you made in step 6. It should now look like a long rectangle with two flaps.

8. With the flaps closed toward the center, fold the top of the rectangle ½ inch down and the bottom of the rectangle ½ inch up. Now fold the bottom of the rectangle up to meet the bottom edge of the flap at the top (see diagram d).

9. Tuck the folded edge of the flaps from the stand into the corners you folded on the back of the frame in step 5 (see diagram e). Each of these corners has a slit down the middle from the folds you made, so you will be able to

tuck in flaps at the top and bottom of the stand at this slit. Tape the stand in place.

10. Cut your photo into a 4-inch x 4-inch square. Insert the photo into the front of the frame by sliding it under the folded space. You can add a piece of clear plastic from a food container or toy packaging to protect your photo. Just cut the plastic into a 4-inch x 4-inch square, and slide it into the frame after you've inserted your photo. Now you're done with this picture-perfect project!

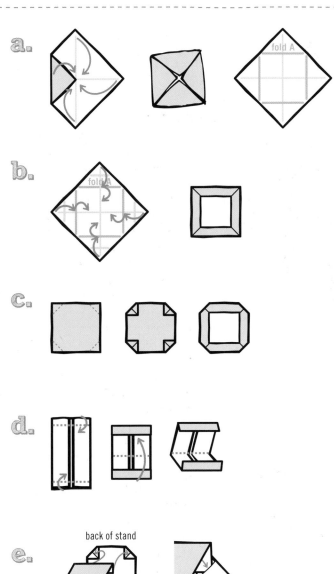

a.

fold A

b.

fold A

c.

d.

e.

back of stand

close-up

stand

2. Paper Box

MATERIALS:
- 2 magazine pages
- scissors
- ruler

INSTRUCTIONS:

1. Measure and cut one sheet of magazine paper into an 8-inch x 8-inch square. Cut another into a 7½-inch x 7½-inch square. The larger square will become the box top.

2. Start with the 8-inch square and fold it in half diagonally. The paper should now look like a triangle. Then fold the triangle in half again straight across the middle. Open the paper flat.

3. Take all four corners of the paper and fold them into the center (see diagram a on the opposite page). Open the piece of paper back up.

4. Fold the left corner of the square piece of paper so the point touches the farthest fold line opposite from it. Unfold it. Repeat this step for each of the other three corners. When you're done, you'll have a square piece of paper that has fold lines that look like the picture in diagram b. It will look like a grid, which will help you in the next step.

5. Place the paper flat so it looks like a diamond. Make two cuts at the top your paper, two squares deep and two squares apart. Make the same cuts at the bottom (see diagram c).

6. Fold the left corner to the first fold line and again to the next fold line. Fold the right corner in the same way. Once you've done this, fold both sides up toward you so they stand vertically. You now have the two walls of your box (see diagram d).

7. Fold in the cut ends of each of the vertical sides to create the other two walls of your box. You should now have what look like the beginnings of a box with two pointed flaps on either side (see diagram e).

8. Take the end of one of the pointed flaps and fold it up twice. The first time should be to the first fold line. On the second fold, bring the flap up and over the wall of your box (see diagram f). Do the same thing with the other pointed flap. Crease the edges of the paper to keep it in place.

9. With your other piece of paper, repeat instructions 2–8 to make the box top. Now you have a perfect box! Use it to store knickknacks, odds and ends, and bric-a-brac. If you have a gift to give, you can use wrapping paper instead of magazine pages to make a gift box that's totally reusable. Once you have the hang of making these boxes, you can create them in other sizes. Just scale the dimensions up or down. Don't let the size keep you boxed in!

You can add a piece of tape to the inside corners if they start to unfold.

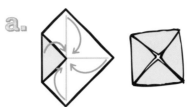
a.
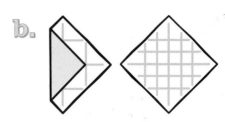
b.

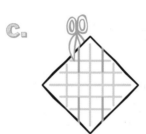
c.
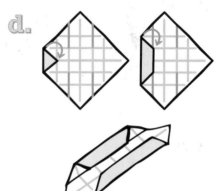
d.

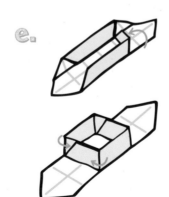
e.
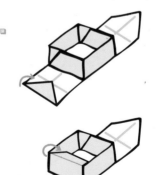
f.

3. Scrap Paper Gift Wrap & Bow

MATERIALS FOR GIFT WRAP:
- sheet music or other scrap paper
- scissors
- tape

INSTRUCTIONS FOR GIFT WRAP:
1. Place the sheet music facedown on a flat surface.

2. Place the gift in the center of the paper.

3. Wrap one side of the paper up and around the gift. Wrap the other side of the paper up and around the gift and tape it to the other side of the paper.

4. Fold the open edges into points by bringing the short sides flat against the gift. The paper will naturally make points at the top and the bottom of the gift. Fold the top point down. Then, fold the bottom point up and over the gift. Tape the bottom point down. Do this on the other side, too.

MATERIALS FOR BOW:
- magazine or other scrap paper
- scissors
- ruler
- stapler
- glue

INSTRUCTIONS FOR BOW:
1. Cut four strips of paper, each 8½ inches long by ¾ inches wide. Then, cut one strip of paper, 4 inches long by ¾ inches wide.

2. Make loops out of two of the long strips of paper and glue them closed. Stack the two loops, one on top of the other, in an "X" shape. Staple these two pieces together in the center (see diagram a).

3. Repeat step 2 for the other two long strips of paper.

4. Take the short strip of paper and make another loop. Stack it on top of the loops from steps 2 and 3. Staple all of these pieces together (see diagram b).

5. Attach the bow to the gift with a little glue. This project is all wrapped up!

a.
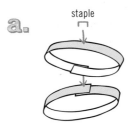
staple

b.
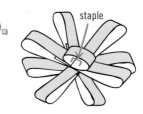
staple

4. Newspaper Gift Bag

MATERIALS:

- thin cardboard (from a cereal or similar box)
- newspaper
- scissors
- tape
- ruler
- paper hole punch
- ribbon or cording for the handle
- pencil
- boxes of different sizes

INSTRUCTIONS:

1. Choose a box the size and shape you want your finished gift bag to be. A closed cereal box works great.

2. Layer two sheets of newspaper on top of each other. Wrap the box just like you're wrapping a present but only around the sides and one end. Make sure you don't tape the paper to the box.

   First, wrap the paper around the box lengthwise. Tape one edge of the paper where it overlaps the other. Then fold in the edges at one end of the box and tape these flaps together. Leave the other end open. This will be the top of your gift bag. Trim the paper with scissors if it hangs more than one inch over the top edge. At the bottom edge of your bag, pinch along the four corner edges of the box to crease the paper.

   You can use just one sheet of paper. But, two sheets will make a stronger, sturdier bag.

3. Slide the box out of the paper and set it aside. Pinch the bag again along the creases to sharpen all the edges.

4. Measure the bottom of your bag. Cut a piece of thin cardboard from your cereal box to be the same size as the bottom of your bag. Place the piece of cardboard into the bottom of your bag for extra strength and form.

5. Place the bag on one of the wider sides so you can flatten it. Fold the bottom of the bag up and, at the same time, press the two shorter sides inward with your fingers. Make a crease where you've folded up the bottom. Make another crease lengthwise down the middle of the shorter sides. This will give you neat fold lines to flatten the bag for storage.

15

6. Measure and cut two strips of cardboard as long as the wider sides of your bag and about 1 inch tall. Fold the top of your bag down about 1 inch and insert the piece of cardboard inside each fold on the wider sides.

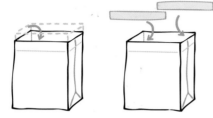

7. Punch two holes through the paper and cardboard on both of the wider sides of the bag. The holes should be about ½ inch down from the top and about 1 inch inward from each edge. Make sure they are lined up with the holes on the other side.

You can also use recycled shoelaces or strips of old T-shirts for the handles.

8. Make two handles for your bag. Tie a double knot in one end of your ribbon. Thread the unknotted end of the ribbon through the hole from the inside of the bag. Thread the ribbon back through the front of the other hole and tie the other end in a double knot on the inside of the bag. Repeat this step on the other side of your bag. These newsworthy gift bags are sure to make headlines!

5. Magazine Envelopes

MATERIALS:
- magazine pages
- ruler
- scissors
- pencil
- tape
- white mailing labels

Decorative paper goes through the mail just fine. But avoid using paper that's printed with lots of letters and numbers. It will make the address harder to read.

INSTRUCTIONS:

1. Measure and cut an 8-inch x 8-inch square out of a magazine page. Measure and mark the very center of the square with pencil (see diagram a on the opposite page).

2. Lay your square on a flat surface in a diamond shape. Fold the bottom corner up and 1 inch past the center mark. Make a crease at the bottom (see diagram b).

3. Fold the left and right corners in to meet the center mark. The bottom corner will peek out a little bit under them. Secure the two sides with pieces of tape (see diagram c).

4. Fold down the top flap about 1½ inches past the center mark. Flip the paper over and place a return-address label and a mailing label on the front. Stuff your envelope, write the correct address on the mailing label, and seal the envelope with tape. Don't forget to add a postage stamp. Once you subscribe to these magazine mail-ables, you won't want to use new envelopes ever again!

Pair this project up with #6 Magazine Greeting Cards to create matching cards and envelopes.

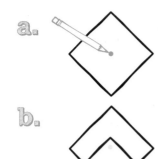

a.

b.

c.

6. Magazine Greeting Cards

SKILL LEVEL:
EASY

TIME: 🕐

MATERIALS:
- sheet of 8-1/2 x 11-inch blank paper
- ruler
- magazine pages
- scissors
- tape

INSTRUCTIONS:

1. Measure and cut the sheet of paper across so it is ½ the original size, or 8½ inches x 5½ inches.

2. Fold the piece of paper in half. It will measure 4½ inches wide x 5½ inches tall.

3. Decorate the card with pictures and letters cut out of magazines. Choose words and images that will help celebrate a special occasion.

4. Stuff your new card into the project **#5 Magazine Envelopes**, and mail away!

7. Paper Tube Organizer

SKILL LEVEL:
EASY

TIME:

MATERIALS:

- 3 cardboard toilet-paper tubes
- construction paper
- scissors
- 3 plastic lids (for example, milk jug lids), each about 1¾ inches in diameter (measured across the middle)
- ruler
- clear tape
- decorative paper
- masking tape
- markers, stickers, or crayons

INSTRUCTIONS:

1. Cut the tops off two of the cardboard tubes to make them different heights. Leave the third alone.

2. Start with one tube and cut a piece of paper to match the height of that tube. Roll up the paper lengthwise and stick it inside. The paper should unroll a little and secure itself in the tube. Place a piece of clear tape over the top of one of the milk jug lids. The tape should stick out about 1 inch on each side of the lid. Place the lid inside the top end of the tube and tape it in place on each side (see diagram a). Repeat this step for the other two tubes. Flip the tubes over so the milk jug lids are now the bottoms.

3. Start with one tube and cut a piece of decorative paper to match the height of that tube. This time cover the tube with paper and place masking tape along the top and bottom edges to secure the paper in place (see diagram b). Repeat this step for the other two tubes. Decorate the three tubes even more with stickers, markers, and anything else you can find.

4. Once the tubes are fully decorated, arrange all three of them in a bundle on a flat surface, lid end down. Mark a line where each one touches the other one, then glue along that line. Attach all three tubes together and let them dry. If you like, you can tie a string around all three tubes to secure them even more. Fill up your new organizer with desk supplies and you'll be on a recycling "roll."

a.

b.

18

8. Oatmeal Container Tote

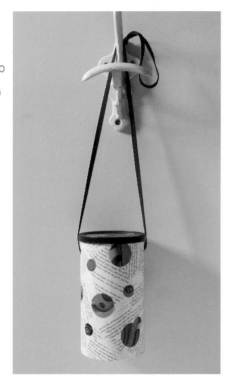

MATERIALS:
- oatmeal container
- paper hole punch
- 12-inch cord or shoelace for the handle
- decorative paper
- 3 tablespoons white glue
- 3 tablespoons water
- plastic container for mixing
- paintbrush

INSTRUCTIONS:

1. Start with a clean, empty oatmeal container. Punch two holes in the container directly across from each other and 1 inch from the top.

2. Cut pieces of decorative paper and glue them onto the container. Arrange the pieces as you wish. You may cover the whole container. Then make a paste by mixing the glue and the water in a container with a paintbrush. Paint a thin coat of the glue/water mixture onto the container with your paintbrush. Make sure the liquid is evenly spread all over. This process is called decoupage. It seals your project with a nice glaze and makes it last longer. Let the container fully dry.

3. Insert the cord handle into the holes you punched in step 1. The ends of the cord should be on the inside of the container. Tie off each end in a double knot. Your oatmeal tote will be part of a well-balanced wardrobe!

Optional: Before inserting the cord handle, put grommets into both holes and secure with a grommet tool for a more finished look.

9. Ice Cream Container Bank

MATERIALS:

- ice cream pint-sized container
- scissors
- pencil
- ruler
- decorative paper
- 3 tablespoons of white glue
- 3 tablespoons of water
- plastic container for mixing
- paintbrush
- markers
- stickers

INSTRUCTIONS:

1. Take the lid of your ice cream container and estimate where you want your money slot to go. It should be right in the center. Mark a thin rectangle with a pencil on the lid of the container. Carefully cut out the slot (see diagram a).

2. Cut decorative strips of paper 1 inch wide and as tall as the container. Glue them around the outside of the container, overlapping them as you glue (see diagram b). Trim off any excess paper.

3. Cut out a piece of decorative paper in the shape and size of the lid. Carefully cut out the slot you cut in step 1. Glue the piece onto the top (see diagram c).

4. Like in project **#8 Oatmeal Container Tote**, make a paste by mixing the glue and the water in a separate container. Paint a think coat of the glue/water mixture onto the container (see diagram d). Make sure the liquid is evenly spread all over. Let the container and lid dry completely.

> You can do this project with other sized containers, too. Just center the slot in the top before you trace and cut it.

5. Place lid back onto the container. Add stickers and use markers to finish decorating it. Here's the scoop—you're ready to start using your new bank!

a.

b.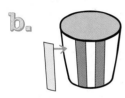

c.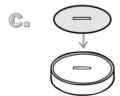

d.

2. Glue pieces of decorative scrap paper onto the switch plate (see diagram a). Arrange the paper in any design you like. Pay special attention to the rectangular hole in the center and the screw holes. If you accidentally cover up these parts, cut those pieces away with your scissors.

a.

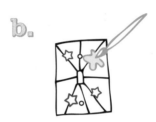

3. In the plastic container, mix a paste of the glue and the water. Paint a thin coat of the paste over the paper on the switch plate (see diagram b). Let it dry fully.

b.

4. With the screwdriver, screw the switch plate back into place on the wall. This is one project that will really light up when it's finished!

MATERIALS:
- screwdriver
- plain switch plate
- scrap paper
- scissors
- 1 tablespoon white glue
- 1 tablespoon water
- plastic container for mixing
- paintbrush

INSTRUCTIONS:
1. Remove the switch plate from the wall with a screwdriver. Before you do this, please ask permission from an **adult** in charge.

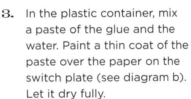

11. Matchbox Jewelry Box

See project **#3 Scrap Paper Gift Wrap & Bow** for how to wrap the back of the stack. Decorate your little matchbox jewelry box with stickers and markers. Your other store-bought jewelry boxes are no "match" for this crafty creation!

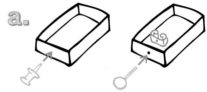

MATERIALS:

- 3 empty matchboxes
- pushpin
- 3 post earrings
- tape
- foil
- scissors

INSTRUCTIONS:

1. With the pushpin, poke a hole into the center of one short side of each matchbox. Insert one earring post into each hole. Secure the earring with the earring backing (see diagram a). These become the drawer handles.

2. Stack the three matchboxes, one on top of the other. Put tape around the long sides of all three boxes so they stay together (see diagram b). Don't tape over the part of the matchbox that slides in and out.

> You can line the insides of the "drawers" with foil for an extra fancy look.

3. Wrap the entire outside of the matchbox stack (except the drawers) with foil and tape it into place (see diagram c).

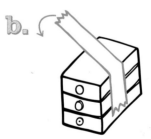

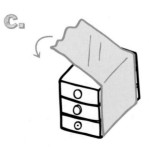

12. Mini Matchbook Journal

You can use recycled scrap paper for your journal—just make sure the blank side is facing up.

MATERIALS:
- opened flattened food box
- scissors
- ruler
- pencil
- scoring tool (like a butter knife or bent paperclip)
- white paper
- scrap cardboard
- pushpin
- stapler

INSTRUCTIONS:

1. Cut a rectangle that is 2½ inches wide by 6½ inches tall out of a food box.

2. On the unprinted side of the food box, mark score lines 2½ inches, 2¾ inches, and 5¾ inches from the top. Line up your ruler with the marks and run your scoring tool along the ruler. Fold inward along your score marks so the outside of the food box shows (see diagram a).

3. Cut ten pieces of white paper to 2½ inches x 2¾ inches. Stack the ten sheets of paper on top of a piece of scrap cardboard. Use your pushpin to poke holes through the paper near the bottom of the page. This will perforate the paper so you can tear sheets out of your journal.

4. Staple through the bottom center of panel 4 (see diagram b). Flip down panel 1 to close your mini journal (see diagram c). Your note-taking skills will be unmatched!

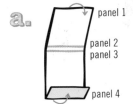

a.
panel 1
panel 2
panel 3
panel 4

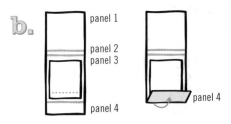

b.
panel 1
panel 2
panel 3
panel 4
panel 4

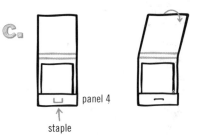

c.
panel 4
staple

13. Record Album Folder

MATERIALS:

- double-record album sleeve with 2 pockets
- scissors
- ruler
- pencil
- elastic cord
- small (⅛-inch) paper hole punch or pushpin
- tape

It's best to use a double-record album sleeve with two pockets. If you only have an album sleeve with a single pocket, cut the album cover along the top and bottom creases and create pockets from 3-inch pieces you will cut from both edges.

INSTRUCTIONS:

1. Ask an **adult** who's in charge to get the OK to use a record album for this project. Open the double record album sleeve and lay it on a flat surface. The album title should be facing up and on the right side. Slip your scissors into the top edge only and cut along the crease.

2. Measure 9½ inches on both sides from the album's spine (which is the exact middle line) and use your ruler to mark a vertical line. Cut through both layers along the lines you marked (see diagram a on the opposite page).

3. Flip the album over so you can see the inside. Measure 5 inches up from the bottom of the album. Using your ruler, mark a straight horizontal line across the entire inside of the album. Cut along this line from both edges. Be careful not to cut through the outside layer (see diagram b). When your scissors reach the center spine, turn and cut in a vertical line along the spine to the top edge. Remove the two parts that were cut loose. Now you have two flaps that you will turn into pockets in step 5.

4. Flip the album back over. On the back cover of the album (the left side), punch a small hole 1 inch down from the top and 1 inch in from the side edge. Do the same thing at the bottom. If you use a pushpin, wiggle it in the hole to make the hole large enough to fit the elastic cord through. Thread the elastic cord through each hole from the outside of the folder. Tie both ends of the cord in double knots on the inside on the folder, leaving the cord slightly loose (see diagram c).

5. From the inside of the folder, tape the pockets in place so they attach to the outside of the folder (see diagram d). Fill your record album folder with your important papers, then close the folder and secure it with the elastic band. This project is sure to be one of your greatest hits!

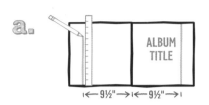

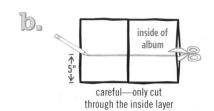

a.

9½" → ← 9½"

9½" → ← 9½"

ALBUM TITLE

ALBUM TITLE

b.

inside of album

5"

careful—only cut through the inside layer

c.

poke hole

ALBUM TITLE

tie double knot

ALBUM TITLE

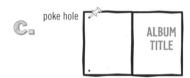

d.

inside of album

tape along edge

optional: reinforce with tape along the bottom

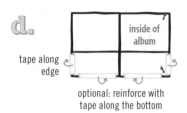

14. Record Album Fan

You can also use cereal boxes or any other type of cardboard for this project.

MATERIALS:

- record album sleeve
- scissors
- ruler
- pencil
- paper hole punch
- bolt and wing nut

bolt

wing nut

- tape measure
- 40-inch long ribbon
- paper clips or binder clips

INSTRUCTIONS:

1. Ask an **adult** who's in charge to get the OK to use a record album for this project. Cut along the top and bottom seams of the album sleeve so you have one large, flat rectangular piece of cardboard.

2. Cut out a rectangle that is 8 inches x 1 inch. Then cut that piece so the bottom tapers to ½ inch wide (see diagram a on the next page). Punch a hole at the top and bottom of the piece (see diagram b).

3. Use the cut piece to trace 11 more pieces onto the cardboard. To maximize the number of pieces you can get from one record album sleeve, trace pieces right next to each other, alternating the direction each time (see diagram a). Cut out the pieces and punch holes in the top and bottom of each one. Make sure the holes are in the same spots on each piece so they will line up properly. You should have 12 pieces total.

4. Stack the pieces, designed side up, on top of each other. Insert the bolt through the front at the narrow end and secure at the back with the wing nut (see diagram c).

5. Fasten the tops of each fan "blades" together with the ribbon. Work from the rear piece going forward to fasten the tops of each fan "blade" together. Starting with the rear piece (which we'll call blade 1), insert the ribbon from front to back through the top hole. Tie a loop in the ribbon around the left side of the blade, double knot it, and pull it snug (see diagram d).

6. Spread blades 1 and 2 so they're resting next to each other. Thread the ribbon through blade 2 from back to front. Then loop the ribbon around front and to the left side of blade 2. Again, thread the ribbon back through blade 2, from back to front. Pull the ribbon just enough so that the right corner of blade 1 and left corner of blade 2 overlap slightly. Secure blade 1 and blade 2, and the ribbon loop, with a binder clip or paper clip (see diagram e).

7. Repeat step 6 with the rest of the blades until you've connected them all except for the last one. When you get to the last blade, thread the ribbon through the hole from back to front. Then, loop the ribbon to the left and all the way around the back of the blade. Then, bring the ribbon around the right side of the blade and to the front. Thread the ribbon through the hole from front to back. Double knot and trim away extra ribbon (see diagram f). You'll stay cool with this chart-topping fan!

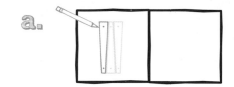

a.

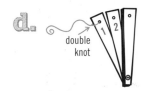

b.   c.

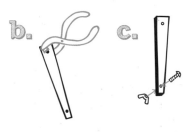

d.

double knot

1 2

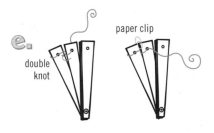

e.

double knot   paper clip

f.

double knot

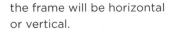
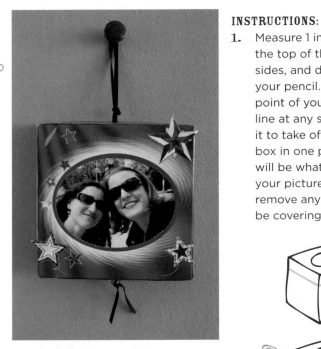

MATERIALS:
- tissue box cube
- ruler
- pencil
- scissors
- markers
- stickers
- clear plastic
- photo
- tape
- paper hole punch
- 18-inch long ribbon
- bead

INSTRUCTIONS:

1. Measure 1 inch down from the top of the box on all four sides, and draw a line with your pencil. Carefully push the point of your scissors into this line at any spot and cut along it to take off the top of the box in one piece. The top will be what you use to make your picture frame. Carefully remove any plastic that might be covering the opening.

2. Decorate the tissue box top with markers and stickers. Keep in mind the size and direction of your photo and make sure it will show in the tissue box hole. Also, decide whether the opening of the frame will be horizontal or vertical.

3. Flip the tissue box frame over so it is face-down on a table. Cut a piece of clear, flat plastic into a 4-inch x 4-inch square. Place the piece of plastic over the oval opening of the frame. Then place your photo, face out, behind the plastic. Tape both the plastic and photo into place.

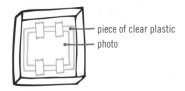

piece of clear plastic
photo

Good places to find clear, flat plastic are on food containers and toy packaging.

4. Use your ruler to figure out the exact center point of the top and bottom of the picture frame. Make a mark at each center point. Punch holes in the spots you've marked on the bottom and top of the frame.

5. Fold the ribbon in half and insert the folded part through the hole in the bottom of the frame. Then thread it through the top hole. Tie a knot in the ribbon about 1½ inches from the top of the ribbon. Your frame will hang from this loop when you're done.

6. Thread the two strands of ribbon at the bottom of frame through a bead. Tie the two bottom ends of the ribbon in a double knot. This picture-perfect tissue frame won't blow your budget!

knot

16. Magazine Travel Bingo

SKILL LEVEL:
HARD

TIME: 🕐🕐🕐

MATERIALS:
- thin scrap cardboard (from a cereal box or similar)
- ruler
- pencil
- markers
- pushpin
- scissors
- magazines
- ball point pen
- tape

INSTRUCTIONS:

1. Cut your cardboard into two 8-inch x 8-inch squares. Set one aside for step 7.

2. Draw a border around the square ½ inch away from the edge. Inside the border, make marks along the top starting at 1 inch, then ½ inch, then

You can go over these lines with a butter knife or open paper clip, making "score marks" that will make cutting and folding easier.

1 inch, alternating this way until you have a line of marks along the top, bottom, and side borders. Use your ruler to draw straight lines connecting these marks. This will create a grid of 1-inch square blocks divided by ½-inch spaces. Decorate each square block with a star or other fun design (see diagram a on the opposite page).

3. Cut along each 1-inch horizontal line from border to

border. Do not cut through the ½-inch spaces between the blocks or the ½-inch border you made in step 2. To make this cutting easier, poke a hole in the cardboard with a pushpin to start each cut line (see diagram b).

4. Cut a vertical line down the center of each square block. The result should look like a 5 x 5 grid of windows with shutters. Fold the "shutters" outward to get a straight crease (see diagram c). Place a ruler along each fold line to make it easier to fold neatly. Then place them back down again.

5. Find magazine images of things you might see from the car on a road trip. Cut each picture into a square about 1¼-inch x 1¼-inch. You'll need 25 of these images to fill the game board. Below, you'll find a list of some ideas for images to use. If you can't find an image you can always draw it or write the name of it behind the shutters. Mark the center space as a FREE space.

6. Tape the square images, face out, to the back of your board, one per square (see diagram d).

7. Take the second 8-inch x 8-inch cardboard square you made in step 1 and tape it to the back of the square with the windows and shutters. This will make your game board sturdy.

8. Repeat these instructions and make as many game boards as you'd like. You'll have tons of ReMade fun on your next road trip.

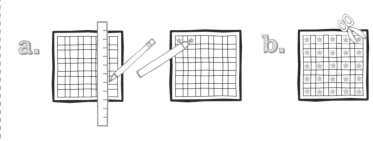

a. b.

c.

d.

back of board

17. Wrapping Paper Basket

MATERIALS:
- used wrapping paper
- ruler
- pencil
- scissors
- string
- clothespins or binder clips
- toothpicks or paintbrush

INSTRUCTIONS:

1. Cut your wrapping paper into 28 strips that are 2 inches wide by 24 inches long.

2. Fold one strip half lengthwise with the printed side facing out. Open it back up. Fold both sides of the strip in toward the center fold line.

3. Fold the strip in half lengthwise again so that you have a ½ inch wide, neatly folded strip. Repeat steps 2 and 3 for all the strips of paper.

4. On a flat surface, arrange seven strips vertically, with the open edge facing to the right. Then, to the right of those seven strips, arrange another seven strips with the open edge facing to the left (see diagram a on the opposite page). The two strips in the center will have their openings facing each other. With your strips laid out vertically, place a ruler about ⅓ inch down from the top of the strips. This will help hold the vertical strips in place while you weave in the horizontal strips. You can weigh the ruler down with something to keep it from moving as you work.

5. Next to the vertical arrangement, lay out strips horizontally just as you did in step 4 vertically. The two center strips will be facing each other, and all other strips will be open toward the center (see diagram b).

6. Weave the horizontal strips in order from top to bottom in your arrangement. Start on the left side of your vertical strip arrangement and weave your horizontal strip over the first strip, and then under the second strip. Continue doing this over and under until you get through all the vertical strips. Next, start weaving the next horizontal strip. Start this next strip under the first vertical strip, and then over the second strip. You can move the ruler once you don't need it to hold the strips in place. Finish doing this for all the horizontal strips you've laid out. You'll have a woven square. Now, shift all the strips toward the middle of the square so there are no gaps in between (see diagram c).

7. Cut a piece of string 70 inches long. Tie the middle of the string around one of your end strips at the corner of the woven square. Then weave one end of the string over and under each strip along the entire perimeter of your square (see diagram d on page 32). Start another weave with the

other end of the string, doing the opposite under and over from the first weave. This keeps the bottom of the basket secure while you work on the sides.

8. The next step is to fold a diamond shape within the woven square that you just made. Take a ruler and place it at a 45-degree angle. Count seven strips in toward the middle of the square and seven strips down from the top of the square (see diagram e). This is where you will place the edge of your ruler. Fold the woven square up at the edge of the ruler. Repeat this for the other 3 sides. This forms the base of your basket.

9. Working on one side of the basket at a time, cross the two center strips. Then cross the second-from-center pieces across the center and over each other. Weave them through the center strips you just folded. Then fold the third-from-center pieces and cross them, weaving them through the other folded strips (see diagram f). Continue doing this until you have a woven diamond shape on each side of the square. The original corners of your square will bend up as you tighten the weaves against each other. The

diamond you folded in step 8 will become the new base of your basket. While you work, tighten the weaves as you go along. It helps to clip the strips you've already woven with clothespins or binder clips and focus working on one area at a time. Continue weaving around the entire basket until it starts to take shape and you've woven all the strips in the manner described in this step.

10. Now, working around the basket, tighten your weaves one last time from the bottom of the basket to the top. You can slip a toothpick or the end of a paintbrush under each weave to help pull it snug. This can be tedious, but it will help make the final product nice and strong, so be patient!

11. Finish off the edges of the basket. Start with one side of the basket and fold the center strip over the last strip that it crossed. Then, weave this folded strip in and out of three or four woven strips on the inside of the basket. Once it's secure (see diagram g), trim the strip with your scissors. Repeat this for the rest of the strips on the basket. This gift wrap basket could make a great gift itself!

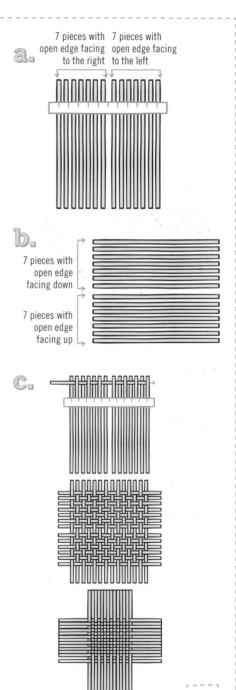

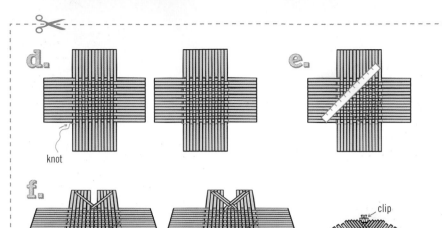

d.

knot

e.

You can also vary the number of strips you use: more than 28 will make a really big basket, and fewer than 28 will make a smaller version. Just be sure to use an even number.

f.

clip

g.

inside of basket

inside of basket

18. Cereal Box Magazine Holder

SKILL LEVEL:
EASY
TIME:

MATERIALS:
- cereal box
- pencil
- ruler
- scissors

INSTRUCTIONS:

1. Start with an empty cereal box, one of its narrow sides facing you. Mark a straight line across this side of the box about 3½ inches from the bottom.

2. Lay the box down so the line you drew is on the right. Then mark an angled line across the front wide side of the box starting from the line you drew to the top left corner. Flip the box over and do the same thing on the other wide side of the box, starting from the line you drew and ending at the top right corner.

3. Open the top flap of the box and start cutting along the lines you marked, beginning at the top right corner. When you're done, stand up your magazine holder and fill it up with your favorite magazines.

19. Box Office Desk

MATERIALS:
- large corrugated cardboard box
- pencil
- scissors
- markers and stickers

INSTRUCTIONS:

1. Open the box and cut off the flaps.

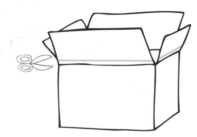

2. Draw a large "U" on one side of the open box, starting your line at the top corner 3 inches in from the side, and ending on the other corner 3 inches in from the side. The bottom of the "U" should be 4 inches from the long side of the box. Cut along the line you drew.

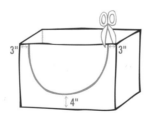

A good place to find a large cardboard box is at an adult's office or at a local grocery store. Large boxes are usually used to package items from computer equipment to multiple rolls of toilet paper.

3. Flip your box over. Decorate it with stickers and markers. Add projects like **#7 Paper Tube Organizer, #79 Tin Can Desk Organizer**, and other containers to help organize all of your supplies. Now you have a nifty new office desk. Think outside the box for other fun ways you can ReMake your play-office!

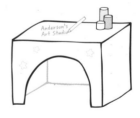

20. Shoebox Storage

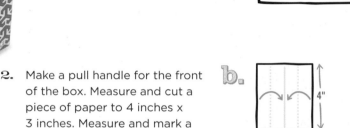

MATERIALS:

- empty shoebox
- stickers
- markers
- decorative paper
- glue
- ruler
- packing tape

INSTRUCTIONS:

1. Take a shoebox and decorate the outside with markers, stickers, and decorative paper (see diagram a). You can attach the decorative paper to the box and use the decoupage technique you used in project **#8 Oatmeal Container Tote**. Attach a photo of your choice to the lid using stickers on the corners.

2. Make a pull handle for the front of the box. Measure and cut a piece of paper to 4 inches x 3 inches. Measure and mark a line lengthwise down the very center of the piece of paper. Fold along this line, then open the paper. Fold the edges lengthwise into this center line and tape them shut. Loop the paper around and tape it inside one of the long edges of the box lid (see diagram b).

3. Place the lid back on the box. Tape the lid to the box with packing tape along the long side opposite from the pull handle. Open up your box and tape the lid to the box from the inside also, on the same side (see diagram c). This will make a reinforced hinge. Start your new storage box on the right "foot" by putting photos and other mementos inside.

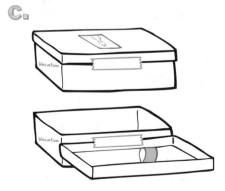

21. Greeting Card Ornament

MATERIALS:
- old greeting cards (**ask permission** to cut these up)
- ruler
- bottle cap
- pencil
- scissors
- glue
- paper clips or binder clips
- paper hole punch or pushpin
- 12 inches of ribbon

INSTRUCTIONS:

1. Find something circular that is about 1 ½ inches in diameter, like a bottle cap or a spool of thread. Place this circle on top of the greeting card. Trace around the edges and cut. Repeat this until you have 21 circular pieces of greeting card.

2. Take one circular piece to make into a triangle template. Draw a triangle in the middle of the circle. All of the sides of the triangle should be equal and each point of the triangle should hit the edge of the circle. Cut along these lines. Take the triangle template and center it on top of one circle. Trace around the edge and fold a valley fold along the three lines (see diagram a on the next page). Repeat this for all the other pieces.

3. On a table, arrange the 20 pieces into two clusters of five pieces and one row of ten pieces, as shown in diagram b.

4. Start with one of the five-piece clusters. Take one piece and put a little glue along one of the inside folded edges. Stick that edge to the edge of the piece right next to it in the cluster. Secure these edges with a clip (see diagram c). Repeat this for the rest of the cluster until the five pieces are connected to each other by their edges. Each piece will be connected on two sides to two different pieces. Repeat this for the other cluster, and let both clusters fully dry.

5. Repeat step 4 for the row of ten pieces (see diagram d). Then connect the first and last piece of the row to create a ring. Let these pieces fully dry.

6. Next, attach both clusters to the ring. Put glue on the top of the folded edges of the ring. Place one of the clusters on top. Secure each folded and glued edge with paper clips and let them dry. Flip this over and repeat the beginning of step 6 for the remaining cluster (see diagram e). Let this fully dry.

7. Using a paper hole punch or a pushpin, punch a hole through one of the flat edges on your ornament. String a piece of ribbon through and tie a knot at the top of the ribbon. Your ornament is ready to hang. Using last year's holiday cards for making next year's ornaments is like ReMaking some cheer.

a.

  b.

c.

d.

e.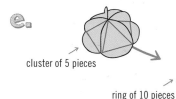

cluster of 5 pieces

ring of 10 pieces

22. Magazine Beads

INSTRUCTIONS:

1. Cut out a rectangle from a magazine page that is 7 inches wide x 1 inch tall. Then cut that piece so the bottom tapers to ¼ inch wide.

2. Place the magazine strips on a flat surface. Lay your toothpick on the wider end and roll the magazine page tightly around the toothpick (see diagram a on the opposite page). Stop rolling before you get to the last inch of the strip.

Once you have the hang of making these beads with magazines, you can try using other types of scrap paper such as old maps, comic books, or last year's calendar.

3. Use another toothpick to spread a small amount of glue on the last inch of the strip. Let the glue set for about 10 seconds, then finish rolling up the strip (see diagram b). If the paper doesn't stick to itself at first, press firmly on the glued

MATERIALS:

- ruler
- magazine pages
- pen
- scissors
- 2 toothpicks
- glue

portion for another 10 seconds. Slide the paper bead you've just made off the toothpick (see diagram c). Let it dry thoroughly.

4. Use your magazine beads as you would use other beads. String them onto stretchy cord to make a bracelet. Add headpins and French hooks to make earrings (see diagram d). String them onto a long thread to make decorative garlands. You can also follow the instructions in project

#28 Straw Wrist Cuff to make a magazine bead cuff. Experiment with different bead sizes by skipping the tapering step in step 1 or by making the bead longer and wider. These ReMade beads can't be beat!

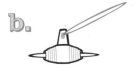

french hook

headpin

23. Gum Wrapper Bracelet

SKILL LEVEL:
EASY
TIME: 🕐

MATERIALS:
- 24 gum or candy wrappers
- scissors
- ruler

Gum wrappers from sticks of gum are usually 2 inches x 2½ inches, so they just need to be cut in half to do this project.

INSTRUCTIONS:

1. Cut your gum or candy wrappers to 1-inch x 2½-inch rectangles. You'll need about 24 wrappers to make a 6-inch bracelet.

2. Take one wrapper and fold it in half lengthwise with the printed side facing out. Open it back up.

3. Fold both sides of the wrapper in to the center fold line.

4. Fold the wrapper in half lengthwise again, so it is now a narrow rectangle ¼ inch wide.

5. Fold the wrapper half widthwise and open it back up.

6. Fold both ends in toward the center fold line.

7. Repeat steps 2 through 6 on another wrapper.

8. Now take your two folded wrappers and fit the two tabs from one piece into the slots of the other (see diagram a).

9. Keep repeating steps 2 through 8 until you've made the bracelet long enough to wrap around your wrist. You'll want to make it slightly larger than your wrist so you can slide it over your hand.

10. To finish your bracelet, follow steps 2 through 5 to fold your last wrapper. Push this piece through the two open ends of your bracelet. Then, tuck the two ends into the center of this piece (see diagram b). Your sweet, new candy wrapper bracelet is ready to slip onto your wrist.

a.

b.

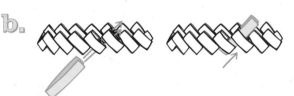

fold and tuck

24. Food Box Postcards

SKILL LEVEL:

EASY

TIME:

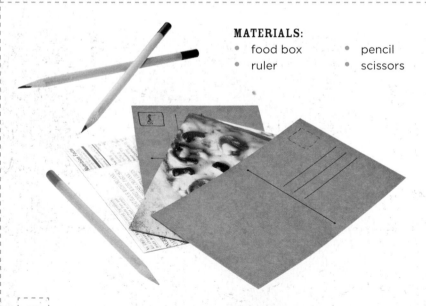

MATERIALS:
- food box
- ruler
- pencil
- scissors

INSTRUCTIONS:

1. Open a food box at the seams and flatten it out.

2. Measure and cut out a 4-inch x 6-inch rectangle.

3. Flip the rectangle to the unprinted side. On the right side of the rectangle, mark lines on the postcard for writing a friend's address. Write a note to your friend on the left side. This postcard contains your recommended daily allowance of fun!

Chapter 2
Fantastic Plastic

25. Chip Bag Bowl

MATERIALS:

- plastic snack-food package (it should be at least 8 inches wide)
- ruler
- pen or marker
- scissors
- tape
- 4 self-adhesive Velcro® fastener tabs

INSTRUCTIONS:

1. Start with a clean plastic snack food package. To clean the inside of a greasy snack food package, wash it with soap and water, then wipe it dry. Flatten out the package and measure the width. Then, place the ruler along the length of the package. Measure and mark lines on both ends to make the length of the wrapper the same size as the width. Cut the wrapper on both ends where you've marked your lines. You will end up with a double-layered square of plastic with two open ends.

2. Place a long piece of tape lengthwise across each of the open ends of the plastic square. Fold the tape over to the back side to seal the ends shut.

3. Mark a border 1½ inches from the edge along all four sides of the snack package. Fold along this line. Unfold and flatten it (see diagram a).

4. Fold up two sides at one corner of the square. Make a diagonal crease from the corner to the folded border to make an upright triangle (see diagram b). Repeat this for all corners, and then unfold them.

5. Unfasten the self-adhesive Velcro® pieces from each other, and stick each one on the inside of each of the corner triangles. Make sure that they'll line up when folded and fastened (see diagram c).

6. Fold up the bowl at the corners and press the Velcro® fasteners together (see diagram d). Fill up your collapsible dish with your favorite snacks!

a.
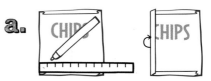

b.

c.
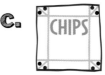

d.

26. Food Wrapper Change Purse

MATERIALS:

- clean, empty food package (at least 8 inches x 6½ inches in size)
- ruler
- washable marker
- scissors
- tape
- 1 self-adhesive Velcro® fastener

INSTRUCTIONS:

1. Start with a clean food package. To clean the inside of a greasy chip bag, wash with soap and water, then wipe dry. Cut a rectangle that is 7 inches wide x 5½ inches tall. Centered inside the rectangle, draw a smaller rectangle that is 4 inches wide x 2½ inches tall. This smaller rectangle will be 1½ inches from all edges.

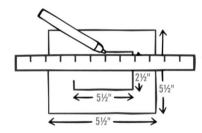

2. On the top edge of the smaller rectangle, draw a semicircle that hits the top of the large rectangle and starts and ends at opposite corners of the small rectangle. Repeat for the other three sides. Cut around these lines.

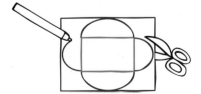

3. Along the lines of the small rectangle, fold all four semicircles towards the center. Your piece should fold up to the size of the small rectangle. Tape the bottom flap and two side flaps in place.

4. Attach one side of the self-adhesive Velcro® fastener on the inside of the top flap. Then, fold down the flap so the other side of the self-adhesive Velcro® fastener attaches to the outside of the bottom flap. Let the Velcro® fastener stick where it naturally landed after you folded the top flap down. Fill your new wallet with all sorts of goodies!

27. Drink Pouch Wallet

SKILL LEVEL: EASY

TIME: 🕐

MATERIALS:
- 2 drink pouches
- scissors
- ruler
- permanent markers
- tape
- 1 self-adhesive Velcro® fastener
- stickers

Save the mini straws from your drink boxes for project #28 Straw Wrist Cuff.

INSTRUCTIONS:

1. Cut off the top of each drink pouch and clean the inside with soap and water.

2. Use the ruler to measure 2½ inches up from the bottom of each pouch, and mark a line across the front of both pouches.

3. Cut two slits down the seams on both sides of the front (designed side) of both pouches. Be careful to cut through only one layer of the pouch. The slits should stop at the marks you made. Then, cut straight across to remove the portion that you just cut (see diagram a).

4. Measure and cut off the top of one pouch so that it is 3½ inches tall. Measure and cut the top off the other pouch so that it is 4½ inches tall. Round the top corners with your scissors (see diagram b).

5. Lay the pouches sealed end to sealed end on a flat surface. The side that you cut into should be face-up. Leave a space between the two ends that is about ⅛ inch, and tape the two pouches together in this position. Make sure you tape both the front and back of the pouches. This will create a hinge so you can fold your wallet (see diagram c).

6. Fold your wallet at the hinge, then fold the longer flap over the top of the shorter flap. Make a crease where you've folded. Then, unfold this flap. Attach the self-adhesive Velcro® fastener to the top of the inside flap of your wallet. Fold the flap down so the other side of the self-adhesive Velcro® fastener attaches to the front of the wallet (see diagram d).

7. Use stickers and permanent markers to decorate your wallet. This simple drink pouch wallet will leave you thirsty for more ReMake It projects!

a.

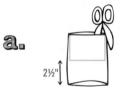

2½"

b.

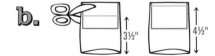

3½"   4½"

c.

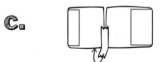

d.

28. Straw Wrist Cuff

MATERIALS:
- 6 skinny, drink-pouch-sized straws
- scissors
- 4 feet of skinny elastic cord
- about 40 small, round beads
- ruler

INSTRUCTIONS:
1. Cut straws into twenty ½-inch pieces.

2. Thread the elastic cord through one piece of straw. Leave an equal amount of cord sticking out of each end.

3. Place a second piece of straw next to the first. Thread a small, round bead through each end of the elastic cord. Make sure the beads you have are big enough so they don't slip inside the straw pieces. Then, thread the top piece of elastic cord from the top down through the second piece of straw. Thread the bottom piece of elastic cord from the bottom up through the same piece of straw.

4. Continue weaving the pieces of straw and beads together in this way until your cuff is long enough to fit around your wrist. Test the length by wrapping it around your wrist. Adjust and add or subtract pieces based on how well it fits. You'll want the cuff to be snug, but not tight.

5. To finish the bracelet, thread one end of the elastic cord up through the very first piece of straw that you threaded in step 2. Tie a double knot at the top of the straw. Pull it tight and make sure the knot doesn't slip back through the straw. Trim the ends of the elastic cord so only ¼ inch of cord remains. Tuck those ends into the straw. Slip on your newly made cuff for a refreshing twist to any outfit!

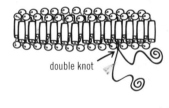

double knot

29. Button Bracelet

MATERIALS:
- buttons
- 30 inches of skinny elastic cord
- scissors

INSTRUCTIONS:

1. Thread the cord through one of the holes of your first button. Leave equal lengths of elastic cord on each side of the button. Tie a double knot on the left side of the button.

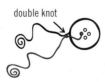

double knot

2. Thread one end of elastic cord through the right hole of the next button. Thread the other end of the elastic cord through the left hole. Then, thread both ends of the elastic cord back through the right buttonhole again. Tie a single knot on the left side of the button.

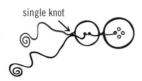

single knot

3. Continue weaving the elastic cord through the buttons in this way, until you have a bracelet-sized row of buttons (measure it against your wrist). Make a loop with the row of buttons, and tie the last button onto the first. This is your button bracelet base strand.

double knot

4. Use one end of the elastic cord to attach the next row of buttons. Thread the end of the elastic cord from the front of the button on the base strand through a new button back to front. Then, thread the cord through the new button, front to back, and back through the next button on the base strand.

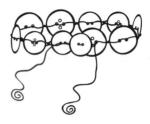

5. Keep repeating this, adding new buttons on top of the base strand of buttons. You'll soon have a new row of buttons overlapping the places where the base row of buttons touches each other. You can continue doing this for a third or even a fourth layer of buttons, depending on how much elastic cord you have left. When the bracelet is as button-filled as you want, tie both ends of the elastic cord in a tight double knot, and trim the extra cord. Now, you have a bracelet that's as cute as a button!

30. Plastic Butter Tub Buttons

MATERIALS:

- plastic containers
- pen
- scissors
- piece of scrap cardboard
- pushpin or small paper hole punch
- needle and thread

INSTRUCTIONS:

1. Collect and clean old plastic containers. The ones that work best are butter tubs, yogurt cups, and sour cream containers. They're colorful and easy to cut with scissors. If you're using round containers, cut down one side and around the bottom, so that you have flat pieces of plastic to work with (see diagram a).

a.

2. Draw simple shapes like circles, hearts, and stars onto the plastic and cut them out (see diagram b). Make sure your shapes are the right size for the buttonholes they will go through. If they are just decorative buttons, they can be any size you want.

b.

You can also make earrings from these cut shapes. Poke the holes where desired and thread an earring wire through the hole.

3. Place one of your cutout plastic pieces on a piece of scrap cardboard. Use a pushpin to poke two holes near the center (see diagram c). To widen the holes so that thread can go through, wriggle the pushpin into each one. You can also find a small paper hole punch at a craft store and use that instead.

c. cardboard / plastic

4. Thread a needle, double-knot the thread, and sew your buttons onto a shirt or sweater (see diagram d). This project works great for clothing that's missing a button.

d.

31. Plastic Container Planters

MATERIALS:
- plastic containers
- plastic lids to fit under the container base
- scrap piece of wood
- hammer
- awl or long nail
- medium-sized rocks or Styrofoam® peanuts
- potting soil
- herb seeds or baby plants

INSTRUCTIONS:

1. Select your containers. Containers that work best are solid in color. Choose a lid for each container that fits with room to spare under the base. The lid will collect overflow when you water your plants.

2. Place the scrap piece of wood on a flat surface and place the bottom of your container on the wood. With an **adult's** help, poke a few holes in the bottom of the container with the awl (or nail) and hammer (see diagram a). These holes will allow water to drain.

3. Place several medium-sized rocks or Styrofoam® peanuts in the bottom of the container. This will help the water to drain through the pot. Fill the rest of the container with potting soil until it reaches about ½ inch from the top. Place the lid under the container with the lip facing up (see diagram b).

← lid

4. Put your seeds or baby plants into the soil (see diagram c). Water them, sit back, and watch them grow. These new planters will put fresh herbs at your fingertips. This project goes great with project **#32 Plastic Plant Markers**.

An awl is a pointed tool used for marking surfaces and making small holes.

32. Plastic Plant Markers

MATERIALS:

- plastic container (a yogurt container works well)
- ruler
- pen
- scissors
- paper hole punch
- old, non-working pen
- permanent marker

INSTRUCTIONS:

1. Collect and clean old plastic yogurt containers. Cut down one side and around the bottom, so that you have a flat piece of plastic to work with. Lay it on a flat surface. Measure and cut out a rectangle that is 3½ inches wide x 2½ inches tall.

2. Punch two holes in the rectangle, one at the top and one at the bottom. The holes should be exactly in the center along the top and bottom edges.

3. Use the permanent marker to write the name of the plant on the plastic.

4. Take the old, non-working pen and poke it through the top hole, from front to back, and then through the bottom hole, from back to front. You may need to use the pen to widen the hole a bit. Or you can punch another hole in the original hole to widen it.

5. Stick the plant marker, pen point side down, into the soil next to the plant. Repeat these steps to make as many plant markers as you wish. This project goes great with project **#31 Plastic Container Planters**.

33. Yogurt Tub Luggage Tags

MATERIALS:

- 1 colored plastic yogurt container
- 1 clear plastic container or lid
- ruler
- pen
- scissors
- scrap piece of cardboard
- pushpin
- paper hole punch
- plastic lanyard or needle and embroidery thread
- ribbon

INSTRUCTIONS:

1. Collect and clean old plastic containers. Cut down one side and around the bottom of each, so that you have flat pieces of plastic to work with.

2. Cut a rectangle out of each container that is 5¼ inches wide x 2¾ inches tall. Measure the exact middle point on one of the short sides of the rectangle. Also measure and mark a point 1 inch from the edge on both the top and bottom long edges. From the middle point on the short edge, mark two diagonal lines, one that goes up to the top point and one that goes down to the bottom point. Cut along the lines. Use your scissors to round all the edges of the piece. Repeat this step for the clear piece of plastic.

3. Use the paper hole punch to punch a hole in the center of the pointed sides of both pieces of plastic.

4. Place the clear piece of plastic you've just cut on top of your scrap piece of cardboard. Use a pushpin to poke holes around the edges, Widen each hole a little bit by wriggling the pushpin in each hole.

5. Next, lay the clear piece of plastic on top of the colored piece of plastic. Lay both of them on the scrap piece of cardboard. Repeat steps 3 and 4 using the clear piece of plastic as your template to create the holes on the colored piece of plastic.

6. With both plastic pieces stacked on top of each other, thread a plastic lanyard or a needle and embroidery string through the first hole on either the top or the bottom of the two pieces. Secure the end of the lanyard with a double knot. Using a running stitch, sew the lanyard in and out of all of the holes around the edge. At the other end, tie another double knot to secure the lanyard. Make sure you leave an opening along the pointed edge so you can slip your address inside.

7. Write your name and contact information on a card that's 2 inches wide x 3½ inches long, and insert it under the clear plastic piece, so your writing shows through the "window."

8. Thread a colorful ribbon through the large hole at the top of the tag, and tie the end in a knot. You are ready for take-off!

34. Plastic Bag Fruit Basket

SKILL LEVEL:
MEDIUM
TIME: 🕐🕐🕐

MATERIALS:

- 3 colorful plastic bags
- scissors
- plastic pint fruit basket (like the ones strawberries come in)
- ribbon (optional)

INSTRUCTIONS:

1. Cut the plastic bags into 1-inch wide strips.

THANK YOU

2. Starting at the bottom of the pint basket, weave your plastic strips in and out of the horizontal row of holes. When you've completely filled in the bottom row of holes, tie the two ends of the plastic bag together. Then, trim any extra plastic and tuck the knots inside. Continue weaving the plastic bag strips like this through the holes of the basket until you reach the top. Try alternating colors to make a nice pattern. You can weave ribbons in some rows to add a different texture.

3. To weave the bottom, start by tying the end of one of the plastic strips in a knot around the center spoke. Then, weave the plastic strip in a spiral

motion around the bottom of the basket. To fill in the corners of the bottom, weave back and forth there. This will fill in the empty areas. When you've filled in all the spaces, tie the plastic strip in a knot around the closest spoke. This woven wonder is ready to be filled with your favorite treats!

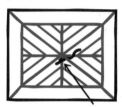

35. Egg Carton Mancala

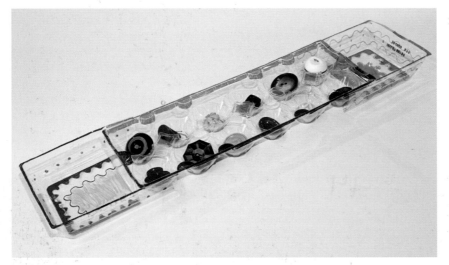

MATERIALS:
- 36 small objects*
- egg carton
- scissors
- ruler
- tape or stapler
- markers and stickers or water-based paint and a brush

*Mancala uses 36 playing pieces. These can be just about anything you want, as long as they're fairly small. You can use buttons, bottle caps, coins, pebbles, or any other small items.

INSTRUCTIONS:

1. Cut off the lid of an egg carton. You will be transforming the lid into two end containers for your Mancala board, so be careful not to tear it.

2. Measure and cut 4½ inches off each end of the lid.

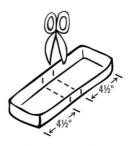

3. On the two ends you just cut off, cut 1½-inch slits into the two creases from the open-ended side. This will create a little flap. Firmly crease and fold up each flap so it is at a right angle to the egg-carton bottom. There should be two little tabs sticking forward on each end piece. Staple or tape the tabs of the two open ends onto the outside of each end of the egg-carton bottom. Now, you have an egg carton with containers on each end.

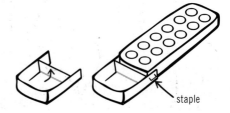

staple

4. Decorate the egg carton and the end containers with markers and stickers. If you are using a paper egg carton, you can paint it and let it dry.

5. Grab your 36 small objects and start playing!

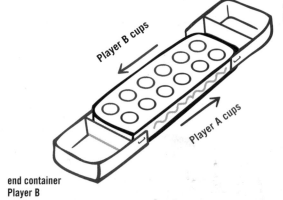

end container Player A

Player B cups

Player A cups

end container Player B

HOW TO PLAY
Egg Carton Mancala:

OBJECT: Collect the most playing pieces in your Mancala end container by the end of the game.

SETUP: Place the Mancala board horizontally between you and your opponent. The game board is split in half, with your closest row of six cups belonging to you. Both players place three pieces in each of his or her six cups.

PLAYING: Decide which player goes first. The first player, say it's you, picks up all three pieces from any one cup on your side of the board. Then, moving counter-clockwise, put one piece in each cup you come to whether it's on your side or your opponent's side. If you come to your own Mancala end container, put one piece in it. If you come to your opponent's Mancala end container, skip it and place the piece in the next cup instead. Take turns with your opponent unless you are eligible for a free turn.

FREE TURN: If your last piece ends up in your own Mancala end container, you get another turn.

At any given point in the game, there may be fewer or more than three pieces in each of your Mancala cups.

CAPTURING: If your last piece ends up in an empty cup on your side of the board, you may capture all the pieces in your opponent's cup directly opposite from yours. Collect these captured pieces and put them in your Mancala end container along with the piece that made the capture. That ends your turn.

ENDING AND WINNING THE GAME: The game ends when all six cups on one player's side of the game board are empty. The player with pieces still remaining on his or her side adds them to his or her Mancala end container. The player whose Mancala end container has the most pieces wins.

36. Plastic Bottle Soap Dish

MATERIALS:
- 1 empty 2-liter plastic bottle
- pushpin
- scissors
- washable marker
- paper hole punch
- (optional) permanent markers and stickers

INSTRUCTIONS:

1. With a washable marker, mark a straight line around the middle of the bottle.

2. Use the pushpin to poke a starter hole a little above the line you drew. With an **adult's supervision**, push the scissors through this hole, and cut down to the line you drew in step 1. Cut along the line, separating the top and the bottom (see diagram a). Be very careful cutting the plastic—the edges can be sharp.

3. Use a washable marker to draw a wavy line between 1 inch and 2 inches from the bottom of the bottle (see diagram b).

4. Using the paper hole punch, make holes at the lowest points of the wavy line (see diagram c). These holes will help guide you while you cut along the line.

5. Cut along the line to make a decorative scalloped edge on your soap dish (see diagram d). Have an **adult** check the edge to make sure it is not too sharp.

6. Decorate the soap dish with permanent markers and stickers (see diagram e). Make sure to decorate the outside of the dish, so the soap won't wash your artwork away!

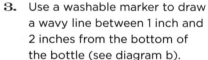

MATERIALS:

- 1 empty plastic bottle (20 oz. or smaller)
- scissors
- pushpin or heavy-duty safety pin
- pen
- paper hole punch
- washable marker
- (optional) permanent markers and stickers

INSTRUCTIONS:

1. Make drainage holes in the bottom (it's easiest to do this while the bottle is still in one piece). Push a pushpin or safety pin through a few places in the lowest part of the bottle (see diagram a). You can enlarge these holes with the point of a pen.

2. With a washable marker, mark a straight line around the bottle about 4 inches from the bottom (see diagram b).

3. Use the pushpin to poke a starter hole near the middle of the bottle. With an **adult's supervision,** push the scissors through this hole and cut down to the line you drew in step 2. Cut along the line, separating the top and the bottom (see diagram c). Be very careful cutting the plastic—the edges can be sharp.

4. Use instructions in steps 3 through 5 of project **#36 Plastic Bottle Soap Dish** to create a scalloped decorative edge around the top of your toothbrush holder. Have an **adult** check the edge to make sure it is not too sharp.

5. Decorate the finished toothbrush holder with permanent markers and stickers (see diagram d). You can also use a paper hole punch to make a fun cutout design. And don't keep your new toothbrush holder all bottled up—there's plenty of room to share with your whole family!

a.

b.

c.

d.

38. Shower Curtain Caddy

MATERIALS:

- 1 used, clean shower curtain
- scissors
- ruler
- washable marker
- binder clips
- straight pins
- sewing machine or needle and thread

INSTRUCTIONS:

1. Using a ruler, mark up your pieces on the shower curtain and cut them out. You'll need two rectangles that are 13½ inches x 10 inches (body of the caddy), two rectangles that are 14 inches x 3 inches (handles), one rectangle that's 12¾ inches x 5 inches (inside divider), and two or three rectangles that are 2 inches x 4½ inches (optional side tabs).

2. Cut a 3-inch square out of both bottom corners of the body pieces (see diagram a on the opposite page). These cuts will become "box corners" later in this project, just like in project **#48 Plastic Bag Pencil Case**.

3. Fold each side of the handle pieces in ½ inch lengthwise. Then, fold each in half lengthwise. Clip both handle pieces together with binder clips or paper clips. Stitch a backstitch in a straight line ½ inch from the open edge (see diagram b). See page 7 for basic sewing tips. When you're sewing on vinyl, use a wide stitch that is between ⅛ inch and ¼ inch wide. Vinyl is sewn a bit differently from fabric. If you're using a sewing machine, sometimes you have to pull the vinyl through the machine, rather than letting the machine feed it through.

4. Hem the top of each body piece by ½ inch. Make sure the "wrong side," that is, the side you don't want to have as the outside of the bag, is facing toward you as you hem. Hem the top and bottom of the center divider piece by ½ inch (see diagram c).

5. Place one of the body pieces on a flat surface with the wrong side facing up. Place the divider piece on top of the body piece 1 inch down from the top. Center it from both edges and pin it in place. Measure the exact center point on the top of the body piece, and stitch a straight line down the center of the two pieces, sewing them together (see diagram d).

6. If you want small outer tabs on your caddy now is the time to hem and then sew them to the body of the shower caddy. Hem the top and bottom by ½ inch. Center the tab piece on the right side (outside) of the body piece. Sew the short ends to the body piece to make a larger strap that can hold a wide hairbrush (see diagram e). If you want your strap to hold smaller items like toothbrushes, just stitch another line down the middle of the strap.

7. Lay one of the body pieces on a flat surface with the wrong side facing up. Take one handle and place each end 3 inches in from the edges of the body piece and 1½ inches down from the top of the body piece. Pin both sides of the handle in place. Stitch an "X" through both ends where it overlaps the body piece (see diagram f). Repeat for the other body piece.

8. With the right sides facing each other, place both body pieces together (match the edges so they are straight).

Sew through both pieces along the bottom and side edges (see diagram g). Make sure you do not sew along the notches or along the divider piece.

9. Now, you will sew up the box corners you cut in step 2. Pick up the body pieces you just sewed together and lay the side seam on top of the bottom seam. Flatten the body of the bag with the side seam laying flat on top of the bottom seam. Stitch a straight line along the box corner, perpendicular to the side and bottom stitches

(see diagram h). Repeat for the other box corner.

10. Flip the caddy inside out so the correct side is facing out. Sew the loose edges of your center dividers to the opposite side of the inside of the caddy directly below the handles (see diagram i). To do this, sew a running stitch through both the fabric of the center dividers and the body of the shower caddy to secure them together. Pack your new caddy with shampoo, conditioner, combs, and ReMake It power!

Lay out the pieces to use the existing grommets at the top of the curtain as drainage holes in the caddy. To do this, use the top edge of the shower curtain as the bottom edge of the caddy's body pieces. Make sure your shower curtain is clean and free of mold.

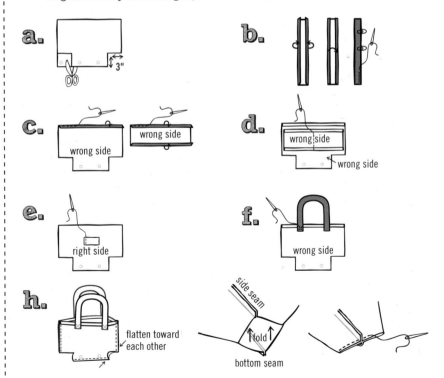

a.

3"

b.

c.

wrong side

wrong side

d.

wrong side

wrong side

e.

right side

f.

wrong side

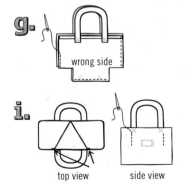

g.

wrong side

h.

flatten toward each other

side seam

fold

bottom seam

i.

top view   side view

39. Lip Balm Keychain

MATERIALS:
- empty lip-balm tube
- needle-nosed (narrow tipped) pliers
- piece of scrap cardboard
- pushpin
- lanyard or yarn, about 3 inches long (lanyard is best, but yarn is OK)
- key ring
- permanent marker, stickers (optional)
- clear packing tape
- small pencil, scrap paper

INSTRUCTIONS:

1. Remove any remaining lip balm and its plastic receptacle from the inside of the tube and discard it. Wash the tube and cap thoroughly in soap and water to remove any grease.

To remove the center spoke from the plastic tube, grab the spoke with a pair of needle-nosed pliers. While holding onto the spoke, twist it until it separates from the container.

2. Place the cap down onto a piece of scrap cardboard. Using your pushpin, make a hole in the center of the cap. Go slowly so you don't crack the cap. Make the hole a little bigger by wriggling the pushpin in the hole (see diagram a).

3. Make a loop with the lanyard and insert it through the hole in the cap. Double knot the bottom piece of the lanyard. You should now have a loop about 1 inch long sticking out the top of the hole. Cut off extra lanyard below the knot so it doesn't stick out of the bottom of the cap. Put a key ring through the loop of lanyard (see diagram b).

4. Decorate the tube with permanent markers or stickers, and then cover it smoothly with clear packing tape so your cool designs don't rub off (see diagram c).

5. Cut some small slips of scrap paper for writing secret messages. Roll them up and insert them into the tube, along with a small pencil (see diagram d). If you and a friend make these together, you can exchange secret messages in your lip-balm key chains—nobody will suspect a thing!

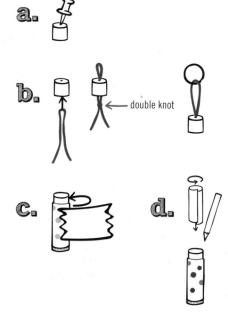

a.

b. ← double knot

c.

d.

40. CD Lampshade

MATERIALS:

- old lamp and lampshade with wire frame
- about 36 CDs*
- thin metal wire (24- or 28-gauge works great)
- wire cutters
- needle-nose pliers (optional)
- ruler
- masking tape

Ask for an *adult's help* when cutting wire and working with sharp objects!

*The number of CDs needed depends on the lamp size. This project calls for twelve strands of three CDs. You can make the lampshade longer or shorter. It's easiest to have an even number of strands.

INSTRUCTIONS:

1. Strip the fabric or paper off the wire frame of the lampshade. (Save this for decorative use in another project like project **#10 Scrap Paper Switch Plate**!)

2. Cut a piece of metal wire to 5½ inches long. Lay two CDs, shiny side-up on top of each other so that only the edge of each CD overlaps. The center holes of the CDs should not overlap at all. With the shiny side up, insert the wire through the center hole of each CD. Flip the two CDs over and twist the two ends of the wire together on the back side of the CDs (see diagram a). Needle-nose pliers will help you do this job neatly. Take another CD, shiny side-up, and overlap its edge with the edge of the CD chain you've started to make. Insert the wire through the center hole of both CDs. Flip the three-CD chain over and twist the two ends of the wire together (see diagram b).

3. Attach your chain of CDs to the lamp frame. Make sure the lamp is unplugged. Cut a 7-inch piece of wire and bend in half. Insert the wire through the center of the top CD of your CD chain. Leave about a ¼-inch gap between the CD chain and the lampshade frame. Attach the CDs to the lampshade frame by wrapping both wire ends around the frame (see diagram c). Cover up the wire ends with masking tape. Repeat step 2 to make more CD chains. Attach them to the lamp frame, allowing them to overlap slightly. Your discs will shine on in their new life!

a. twist

b. 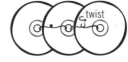 twist

c.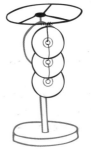

41. Plastic Jug Lampshade

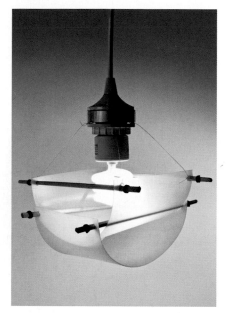

MATERIALS:

- empty gallon-jug of vinegar or distilled water
- ruler
- washable marker
- pushpin
- scissors
- hanging lamp
- paper hole punch
- 2 pairs of chopsticks
- small rubber bands
- fishing line

INSTRUCTIONS:

1. Use your ruler and marker to measure and mark two lines around the plastic container: one will be ½ inch from the bottom and the other will be 5½ inches from the bottom (see diagram a on the opposite page). Then, mark two vertical lines down the sides of the container, directly across from each other.

2. Use the pushpin to poke a starter hole above the area you marked (see diagram b). With an **adult's supervision**, push the scissors through this hole and cut down to the top line you drew in step 1. Cut along this line to take the top off the jug. Near the bottom of the jug, use the pushpin to poke a starter hole below the area you marked. With an **adult's supervision**, push the scissors through this hole and cut up to the bottom line you drew in step 1. Cut along this line to take the bottom off the jug. Cut down the vertical lines you drew in step 1. Then, wash off any marker left.

3. Unscrew the bottom disc from the unplugged hanging lamp, and trace around it onto a leftover piece from the plastic bottle (see diagram c). You will eventually sandwich this piece between the lamp and the screw-on disk. Your lamp will be attached to and hang from this piece. Cut along the lines you marked, but cut the center circle slightly inside the line you marked. To cut the center circle, use the pushpin to poke a starter hole inside the marked area. With an **adult's supervision**, push the scissors through this hole and cut slightly inside this line. Set this piece aside for step 7.

4. Punch a hole into all 4 corners of the plastic panels you cut in step 2 (see diagram d).

5. Take one panel and slide a pair of chopsticks through the holes (see diagram e). Secure the chopsticks by placing rubber bands around each of the ends.

6. Bend the second piece of plastic around the bottom of the first in the opposite direction. Then, thread the remaining two chopsticks into the holes of this plastic panel and secure them at the ends with rubber bands (see diagram f). Make sure the chopsticks in this piece hang across the bottom of the first piece.

7. Cut four 11-inch pieces of fishing line. Tie one end of the lines to each side of the two chopsticks on the top panel. Then, tie the other ends of the four lines to the "donut" shape you cut in step 3. Make sure these are secure double knots

8. Slide the "donut" onto the light fixture. Then, screw on the lamp disc and a light bulb into the fixture (see diagram g). Now, you're ready to plug in your new creation. This lamp will shed some new light on recycling!

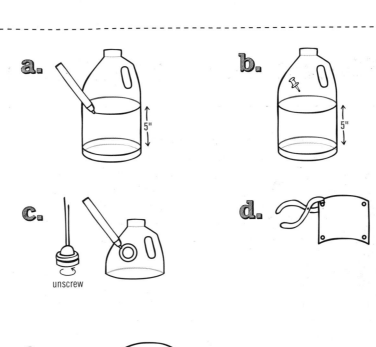

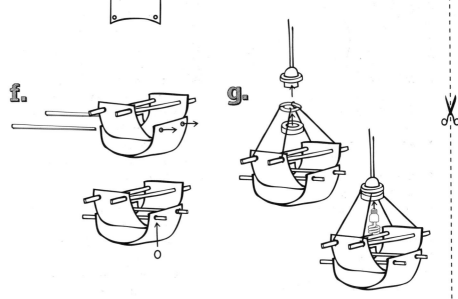

42. CD Trinket Box

MATERIALS:

- thin cardboard (a cereal box or similar)
- ruler
- pencil
- scissors
- clear tape
- CD spindle case (this project uses the outer casing from a CD spindle. You can save the center spindle for project **#43 CD Case Photo Spinner**.)
- permanent markers
- paper hole punch
- 2 CDs
- ribbon or thread

INSTRUCTIONS:

1. Place the cardboard on a flat surface with the blank side up. Use a ruler and pencil to mark and cut three rectangles that are 4¾ inches x 3 inches. Fold each rectangle in half lengthwise so the blank side is on the inside of the fold. This will make a center crease in each piece. Decorate the blank sides with permanent markers and stickers.

2. Stand the rectangles on their ends with the creases touching and the decorated sides facing out. Tape the edge of one side of rectangle 1 to rectangle 2, tape the edge of rectangle 2 to rectangle 3, and tape the edge of rectangle 3 to rectangle 1 (see diagram a). Then, reinforce the three pieces with a little more tape where the edges touch at the top and bottom. This will be your center divider piece. Put this into the plastic shell and move the cardboard around until you have three sections inside that are roughly equal in size.

3. Draw and cut two flowers out of thin cardboard. Punch a hole in the middle of each one.

4. Stack the CDs and cardboard flowers so that the two CDs are in the center, shiny side facing up, with a cardboard flower on both the top and bottom of the stack. Insert both ends of your ribbon up through the holes in the stack. Tie a knot in the bottom of the ribbon. Tie a knot in the top of the ribbon right above the holes so the ribbon doesn't slip through. You will have a loop coming out of the top. This will be your trinket box lid (see diagram b).

5. Decorate the CDs and spindle case with permanent markers and stickers. Fill your trinket box with fun stuff and put the lid on top with ribbon loop facing up. Your old CDs will now be playing a different tune.

> You can decorate with permanent markers before you assemble everything.

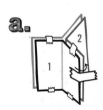
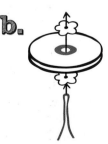

43. CD Case Photo Spinner

MATERIALS:

- 5 CD cases of the same thickness
- 10 photos
- ruler
- pencil
- scissors
- glue
- decorative paper
- stickers, ribbon, scrap paper, etc.
- tape (electrical tape works best)
- 10 old CDs
- CD spindle

INSTRUCTIONS:

1. Carefully remove the black center piece of each CD case by lifting up on one edge until the whole thing moves freely (see diagram a on the next page). This is the center plate that the CD would snap onto.

2. Make ten paper photo displays, one for each side of the CD cases. Trim large photos to 4¾ inches x 4¾ inches, or glue smaller photos onto 4¾-inch x 4¾-inch pieces of decorative paper. Decorate photos with stickers, ribbon, scrap paper, and more. Insert two photo displays, back to back, inside each CD case (see diagram b). The photos will display where the CD liner notes used to be. Close the CD cases.

3. Stack two cases on top of each other. Tape the cases together along their spines so that you can still open the cases when you want to switch the photos. Add another case to the top of the stack and tape it to the one beneath it (see diagram c). Repeat this step until you have a stack of five cases taped together on their spines.

4. Place the case stack upright so it is standing on your work surface with the front facing you. Fan the cases out so they are evenly spaced. They should be touching at the center where the tape is. Insert more tape at the hinge between the two cases that used to be on the top and the bottom of the stack. Then, reinforce all the other hinges with tape (see diagram d).

5. Put your CDs, shiny side up, in the bottom of the CD spindle. The CDs will provide a slippery surface for the photo spinner to spin on. Put the CD case spinner onto the spindle. It's best if the spindle is taller than the CD cases. The center hole where they're taped together should fit perfectly onto the spindle (see diagram e). If the cases don't spin easily, try gently tugging on them so the tape stretches slightly. Place the CD case photo spinner on your desk and admire your work. This project is a great way to display your favorite photos!

This project uses the center spindle from a stack of CDs. You probably saved this piece when you made project #42 CD Trinket Box

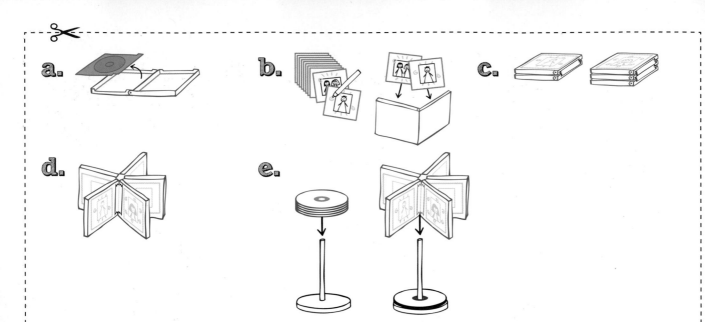

44. Milk Crate Storage Unit

MATERIALS:

- 3 milk crates
- 15 plastic bags
- scissors

If you don't have milk crates, ask a local grocery store if they have any extra.

INSTRUCTIONS:

1. Place the milk crates side by side on a flat surface with the open sides facing up (see diagram a on the opposite page).

2. Cut ¼ inch off the bottom one plastic bag. Then, cut a continuous spiral that is 2 inches wide around the bag. Repeat this with all fifteen plastic bags (see diagram b).

3. Starting from the inside of the top crate, thread both ends of a plastic bag strip through the set of holes that are closest to the second crate. Crisscross both plastic bag ends, and then thread through the pair of holes on the second crate. Tie the plastic bag strip into a double knot on the inside of the second crate and trim the extra plastic (see diagram c). Repeat this step to tie together all crates together on both sides of crates. Make sure all of the crates are tightly tied to each other.

4. Take another plastic bag strip and tie it loosely around the top edge of the top crate (as shown in diagram d). Then, weave the plastic strip in and out of a horizontal line of holes on one side of a crate. When you come to the end of that crate, continue weaving the plastic strip into the next crate and the next crate. Round the corner and continue weaving around the crate, around the next corner, back down the other side, and continue weaving through that one row until you get back to where you started. Untie the knot in the plastic strip and tie the two ends of the plastic strip together in a double knot.

5. Repeat step 4 with the rest of the plastic bags, securing the three crates together on all sides. When they are secure, pick up the crates and place the stack upright with the open side facing out. Now, you have a new set of shelves you made yourself!

If you have an uneven number of holes and your plastic strip ends up on the inside of the crate when it's supposed to weave into the next crate, thread it through a neighboring hole to put it back on the outside of the crate.

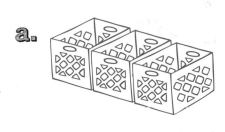

a.

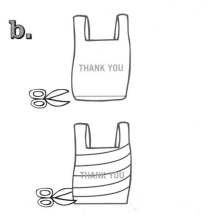

b.

THANK YOU

THANK YOU

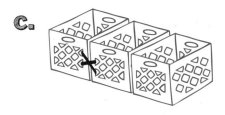

c.

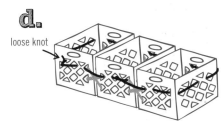

d.

loose knot

45. Record Clock

MATERIALS:

- scrap paper
- pencil
- scissors
- vinyl record
- glue stick or rubber cement
- clock mechanism kit (available at craft-supply stores)
- battery for the clock mechanism
- picture hanging hardware (optional)

INSTRUCTIONS:

1. Cut out the numbers 1 to 12 from your scrap paper (see diagram a). You can use sheet music, newspaper, magazines, junk mail, or whatever you like.

2. Glue the numbers to the front of your record album (see diagram b). It's easiest to start with the numbers 3, 6, 9, and 12 at quarter points around the clock to make sure the numbers are evenly spaced. Then, fill in the other numbers around the clock.

3. Most clock mechanisms you can buy are just a little too large for the hole in the center of the vinyl record. To widen the hole by just a little bit, get an **adult's help** to open a pair of scissors and gently slice off a small bit all the way around the center hole (see diagram c). This is tough, so be careful! Widen the hole until the clock mechanism fits through it.

4. Insert the clock mechanism through the hole in the vinyl record, screw the mechanism into place, add the minute and second hands, and put the battery in place (see diagram d). If you have picture hanging hardware, ask an **adult** to hang the clock in your room. Or, you can set it on your dresser and lean it against the wall. Now, you're ready to rock around the clock!

> Use the vinyl record you took from the album sleeve for project **#13 Record Album Folder** or **#14 Record Album Fan**. Again, make sure it's a record you have permission to ReMake!

a.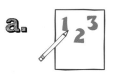

b.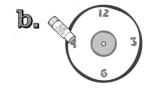

c.

d.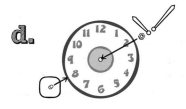

46. Yogurt Cup Desk Organizer

MATERIALS:

- 3 6-oz. yogurt cups
- scissors
- washable marker
- ruler
- paper hole punch
- 7 inches of elastic cord
- 1 straw
- clear tape
- ribbon
- permanent markers (optional)

INSTRUCTIONS:

1. Draw an angled line around one of the containers (see diagram a). Then, cut along the line you drew. Repeat this step for the two other cups.

2. Use your ruler to measure and mark two holes onto the longest side of each cup. These holes will be 1½ inches from the bottom of the cups and ½ inch apart. The marks should be as centered as possible. Punch the holes with a paper hole punch (see diagram b).

3. Starting from the outside of one cup, weave the elastic cord into the left hole and out the right hole. Then, move on to the next cup, weaving the cord into the left hole and out the right one. Then, move on to the next cup, weaving the cord to connect all three cups. Pull the two long pieces of leftover cord up through the center of the three cups and tie the ends into a double knot (see diagram c). Make sure you pull the cord extra tight to keep your cups close together.

4. Measure and cut the straw 5 inches long. Then cut two 1-inch slits opposite each other on the bottom of your straw. Fold up these pieces of straw. Tape the straw flaps into place (see diagram d). Push the straw through the opening at the center of your cups with the taped part resting just under the elastic. This straw is a handle so you can pick up your organizer and move it around.

5. Label and decorate your cups with permanent markers and stickers. Your crafty new creation will help keep your art supplies organized.

You can place your uncut cup inside the cup that you just cut. Use the cut cup as a template for the uncut cup.

a.

b.

c.

d.

47. Plastic Bag Placemat

MATERIALS:

- 5 plastic grocery bags, all the same size
- ruler
- washable marker
- scissors
- wax paper or parchment paper
- iron and ironing board

INSTRUCTIONS:

1. Measure and cut ¼ inch off the bottom of each plastic bag.

2. Turn the bags inside out and stack them on top of each other. Place the stack of bags on one large piece of wax or parchment paper and cover the stack with another large piece of wax or parchment paper (see diagram a). Place this entire stack on the ironing board.

3. With an **adult's supervision**, set your iron to medium heat (the rayon or polyester setting). Do a test: iron one corner of your stack of bags to make sure they don't burn. If they do, turn the iron down. Once you have the correct setting, slowly run the iron over the top of your paper, always keeping the iron moving. Run the iron over the bags several times, but never touch the hot iron directly to the plastic (see diagram b). This technique fuses the plastic bags together so they become a solid piece of plastic. Turn off your iron and set it aside.

4. Take the fused plastic bags out from the layers of wax or parchment paper. Using a ruler, mark and cut out a rectangle on the fused plastic that is 17 inches x 11 inches. Wipe off any marker you can still see. Round the corners with your scissors (see diagram c). Make as many placemats as you'd like. Then, serve on these ReMade placemats at your next meal.

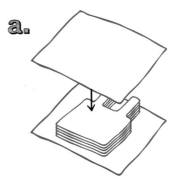
a.

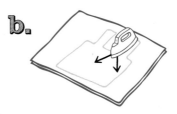
b.

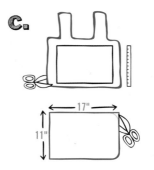
c.

17"

11"

48. Plastic Bag Pencil Case

MATERIALS:
- 5 plastic grocery bags
- ruler
- washable marker
- scissors
- wax paper or parchment paper
- iron and ironing board
- tape or straight pins
- 12-inch zipper
- needle and thread

INSTRUCTIONS:

1. Repeat steps 1 through 3 from project **#47 Plastic Bag Placemat** to fuse the five plastic bags together. Be sure to have an **adult** help you.

2. Using a ruler and washable marker, mark and cut from the fused plastic two 11-inch x 4-inch rectangles. Wipe off any marker you can still see. Then, with the rectangles lying horizontally, mark and cut two

1-inch squares from the bottom corners of each rectangle, just like you did in project **#38 Shower Curtain Caddy**.

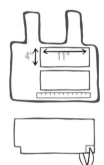

3. Fold the top of each rectangle down ½ inch on the wrong side of the material. Use a little bit of tape or straight pins to help keep the fold in place.

wrong side

4. Lay out your zipper in a straight line on the table. With the folded side facing down, place your rectangle on top of the zipper at the ½-inch fold. Use some tape or straight pins to keep the zipper in place.

Using a needle and thread, stitch the zipper in place using a running stitch. Your stitch should be about ¼ inch from the zipper teeth and centered on the ½-inch fold. Then, repeat this step with the other piece of fused plastic on the other side of the zipper.

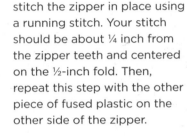

5. Unzip the zipper about halfway. When you're finished sewing the bag, you're going to need this opening to flip your bag inside out. With the right sides of the plastic facing each other, fold your piece in half along the zipper. Sew along the sides and bottom edges making sure not to sew along the cut out notches on the bottom corners.

wrong side

6. Now, you'll sew up the box corners you cut in step 2. Pick up the body pieces you just sewed together and match up the side seam and bottom seam at the notches. Flatten the body of the bag this way, with the side seam laying flat on top of the bottom seam. Stitch a straight line along the box corner, sideways to the side and bottom stitches. Repeat for the other box corner.

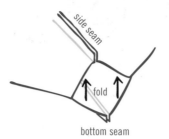

side seam

fold

bottom seam

7. Unzip the entire case and flip it inside out so the correct side is showing. Now you have a zippy, ReMade case for your pencils!

49. Plastic Lid Coasters

SKILL LEVEL:
EASY
TIME: 🕐

MATERIALS:
- plastic lids, clear or opaque, from yogurt and sour cream
- scissors
- permanent markers

- decorative papers
- clear packing tape
- paper hole punch
- hot glue gun (have an **adult** help you!) or rubber cement

INSTRUCTIONS:

1. Find two lids, one slightly smaller than the other. The smaller lid will be the top of your finished coaster.

2. Cut the lip off the edge of the smaller lid. Then, decorate both lids. If the smaller lid is clear, you can decorate the top with permanent markers or make a collage on the underside with decorative papers. Cover the collage with clear packing tape so it doesn't peel off. If the smaller lid is not clear, decorate it by cutting out a fun design with scissors or a paper hole punch.

3. Place the smaller lid inside the larger lid. Glue the two lids together with a dot of hot glue or rubber cement in the center. This plastic lid coaster project just can't be topped!

Chapter 3
Re-Fabbed Fabric

50. Necktie Gadget Case

MATERIALS:

- tie (make sure it's one that an **adult** has given you permission to use!)
- ruler
- fabric scissors
- washable marker or colored pencil
- large button (about 1 inch in diameter)
- needle and thread
- 3-inch thin elastic cord or a ponytail holder

INSTRUCTIONS:

1. Lay the tie on a flat surface, the wrong (seamed) side facing up. With your ruler, measure 12½ inches from the bottom point of the tie. With your washable marker, draw a line across the tie at that point, and cut with scissors. Set the upper, skinny portion of the tie aside for another project, like the handles on project **#61 Comforter Laptop Cover**.

2. With the wrong side of the tie still facing up, make a fold ½ inch down from the cut end. Sew in place with a running stitch (see diagram a).

> Gadgets come in all shapes and sizes. This necktie case will fit most standard-sized cell phones and MP3 players. But you can change the height of this case if your gadget is extra large or small.

3. Flip the tie back over onto its right side. Place a large button about 2 inches down from the sewn edge. Sew this button into place (see diagram b).

4. Flip the tie over so the wrong side is facing up and the pointed end points away from you. Fold up the bottom 4 inches of the tie and stitch up the left and right edges to form a pocket. The button should be on the front and center of this pocket (see diagram c).

5. Tie the ends of a 3-inch piece of thin elastic cord together in a double knot, or use a ponytail holder. Attach this piece to the project by sewing the knotted end of the cord, or any part of the ponytail holder, onto the inside edge of the pointed tip of the tie (see diagram d). The pointed end of the tie is now a top flap to the gadget case and the elastic cord is now the fastener. Flip down the flap and hook the cord over the button to secure your gadget in the necktie holder. This new necktie case will really dress up your favorite gadget!

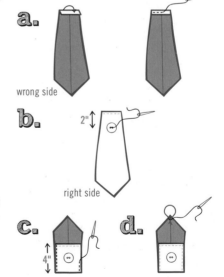

a. wrong side

b. 2" right side

c. 4"

d.

51. Sports Jersey Tote

MATERIALS:
- old sports jersey
- ruler
- fabric scissors
- needle and thread
- 36 inches of ribbon, cut in half
- safety pins

INSTRUCTIONS:

1. Cut through your jersey across the chest. Put aside the portion with the arms and neck hole—you will not need it for this project. Keep the sleeves for project **#52 Shirt Sleeve Bag**. Cut a square 1½ inch x 1½ inch into both corners of the side of the jersey you just cut. Make sure you cut through both the front and back of the shirt.

2. Turn the jersey inside out. Sew a backstitch along the cut edge, but do not sew along the squares you cut out in step 1 (see diagram a). The bottom, hemmed edge of the shirt will become the top of the bag.

3. Now, sew up the box corners you cut in step 1. Pick up the side you just sewed together in step 2 and match up the side seam and bottom seam. Flatten the shirt this way, with the side seam laying flat on top of the bottom seam. Stitch a straight line along the box corner, perpendicular to the side and bottom stitches. Repeat for the other box corner (see diagram b).

4. Rotate your bag so the open end is facing you. It should still be inside out. Pin one end of one piece of ribbon to the wrong side of the open end of your tote, about 2 inches from the edge. Pin the other end of the ribbon to the outside of the open end of your tote, about 2 inches from the edge. Stitch an X onto both ends of the ribbon to secure it in place (see diagram c). Flip the bag around and repeat this step for the other side. You now have two handles.

5. Turn your bag right side out and use your new tote to carry your gym clothes and other essentials. Your new tote will show you're a good sport!

a. wrong side
top of tote

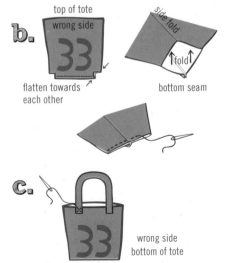

b. top of tote
wrong side
flatten towards each other

side fold
fold
bottom seam

c. wrong side
bottom of tote

52. Shirt Sleeve Bag

SKILL LEVEL: EASY

TIME: 🕐

3. Cut two small slits through both layers of the hem, on opposite ends of the sleeve (see diagram c). Be careful not to cut through the stitched part of the hem.

Optional: instead of using shoelaces for the drawstring, you can use long strips cut from a T-shirt. Simply cut two 18-inch-long pieces from a T-shirt that are ¼ inch wide and pull on the ends so the edges curl. Attach a small safety pin to the end of the T-shirt material to thread it through the hemmed edge.

MATERIALS:
- T-shirt
- fabric scissors
- needle and thread
- two 24-inch shoelaces
- safety pins

INSTRUCTIONS

1. Cut off the sleeve of the T-shirt near the armpit (see diagram a). Or, use the sleeves you've saved from project **#51 Sports Jersey Tote.** The hemmed edge of the sleeve will become the casing for the drawstring.

2. Turn the sleeve inside out and backstitch along the edge that's not hemmed, sewing that edge shut (see diagram b). Flip the sleeve back to right side out.

4. To make the pouch drawstrings, thread the hard tip of one shoelace into one of the holes. Move it through the entire hemmed edge of the sleeve, coming back out through the same hole. Tie both shoelace ends together in a double knot. Repeat this step with the second shoelace, through the hole on the opposite side. Once you've secured the second shoelace with a double knot, pull the shoelaces at both sides away from each other, and you've closed your pouch (see diagram d). Pull this project out of your sleeve whenever you grow out of your old shirts.

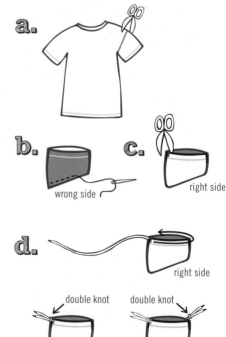

53. Skirt Laundry Bag

SKILL LEVEL:
HARD
TIME:

MATERIALS:
- skirt
- safety pins
- needle and thread
- fabric scissors
- spool of cord
- measuring tape

INSTRUCTIONS:

1. Turn the skirt inside out. If there's a hem around the top of the skirt, cut a ½-inch slit on each side at the seams (see diagram a on the next page). Remove any drawstring or elastic that is already in the skirt. If there is no hem around the top of the skirt, make one. Fold the fabric down 1 inch from the top toward the inside of the skirt. Pin it in place. Then, sew a backstitch along this fold ¾ inch down from the top. Then, cut the two ½-inch slits.

2. Cut two pieces of cord. For an adult-sized skirt, cut 100 inches of cord for each side. For a child-sized skirt, cut 75 inches of cord for each side. Attach a safety pin to one end of the cord. Starting at one of the slits you cut at the waist, thread the cord through the hole and around the entire hem, coming back out through the same hole (see diagram b). Adjust the cord so both ends sticking out are even.

3. Repeat step 2 with the other cord on the other side of the skirt (see diagram c).

4. Pull the two ends of one set of cords through the top of the skirt and lead them all the way down the inside and through the bottom. Pull the cords toward the bottom side edge. Leave ½ inch of cord sticking out through the bottom. Repeat this with the other set of cords (see diagram d). These cords become the drawstring and can be used to carry the bag over your shoulder.

5. Place safety pins along the bottom edge of the skirt every few inches. Sew a backstitch along this edge, closing up the entire bottom of the skirt. Make sure you stitch across the cords, too, so they are held in place (see diagram e). Reinforce, or double up, your stitches at the points where the cords are sticking out. Flip the skirt back to right side. To close your new laundry bag, pull the drawstring cords away from each other at the top. This project will surely help you clean up your wardrobe!

It's best to use a big, flowy skirt to make this Skirt Laundry Bag.

a. wrong side

Another idea: Use a smaller, pencil skirt to make a backpack pouch.

54. Fancy Dress Pillow

MATERIALS:
- dress with a back zipper
- ruler
- washable marker
- fabric scissors
- safety pins
- needle and thread
- pillow stuffing

INSTRUCTIONS:

1. Make sure you ask permission from an **adult** before you cut up a dress. For a 12-inch x 12-inch pillow, you'll need to cut two square panels that are 12¾ inches x 12¾ inches.

2. Make the front panel: Measure a square section 12¾ inches x 12¾ inches, and mark it with your washable marker. With your fabric scissors, cut along the lines you drew. Pick out some features of your dress that you want to display. There might be a fancy bow in the middle, or some buttons, or some rhinestones. Make sure this part is featured on the square you cut out of the dress. Or you can cut an extra piece with the dress's best feature, and use a running

stitch to sew it onto the 12¾-inch x 12¾-inch square.

3. Make the back panel: Out of the back of the dress, measure and cut a square that is 12¾ inches x 12¾ inches. Make sure the zipper is centered vertically, but comes all the way up to the top edge of the square. If you don't have enough fabric, cut additional strips of fabric from the rest of the dress and sew them as needed to make the square.

4. Unzip the zipper about halfway—you'll need this opening so you can turn the pillow back to the right side out after you've sewn the edges. Stack the squares on top of each other with the right sides facing each other. Pin along all edges of the square and use a running stitch to sew straight lines along all edges. As you stitch across the zipper, sew back and forth across the top of the zipper a few times to reinforce the seam.

5. Notch the corners and turn the fabric right side out. Insert fabric scraps, stuffing, or a pillow form and zip your pillow closed. Take a nap and rest from your hard ReMake It work!

You can make recycled pillow stuffing by using small pieces of scrap fabric.

55. Pillowcase Art Supply Holder

MATERIALS:
- pillowcase
- ruler
- washable marker or colored pencil
- fabric scissors
- safety pins
- needle and thread
- 48 inches of ribbon

INSTRUCTIONS:

1. Lay pillowcase horizontally on a flat surface. Measure 15 inches from the bottom of the pillowcase and mark a horizontal line across. Cut along the line with your fabric scissors. Turn the pillowcase inside out and sew a running stitch along the cut edge (see diagram a on the next page).

2. Turn the pillowcase right side out and lay it horizontally on a flat surface. Fold the bottom up 5 inches (see diagram b).

You'll now have a folded pillowcase 10 inches tall. The fold creates the pocket layer.

3. Cut one piece of ribbon 18 inches long and one 30 inches long. Place the ribbon into the large open end of the pillowcase 5 inches down from the top. Let the extra ribbon hang out of the open end of the pillowcase. Carefully insert safety pins along the top edge of the part you folded up in step 2. Make the pins go through the ribbons, too. This will hold the front panel to the larger back panel and keep the ribbons in place (see diagram c).

4. Sew along both side edges of the pillowcase. Make sure to sew the front pocket layer and larger back panel together (see diagram d).

5. Measure 8 inches out from the left edge of the pillowcase and mark a vertical line with the colored pencil on your front layer. Then, measure and mark lines from there in 1-inch sections until you reach the right edge (see diagram e).

6. Stitch through the pillowcase along all the marked lines you made, creating pockets for your art supplies (see diagram f).

7. Insert art supplies into pockets, roll up the case, and tie it shut by wrapping the longer ribbon around the case and tying a bow with the shorter ribbon (see diagram g). Your new case is a great place to put your art supplies to rest!

The wide pocket will fit a notebook or sketchbook and the small pockets will fit pencils, paintbrushes, or scissors. Feel free to change the widths depending on what you want to keep in your holder.

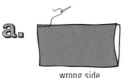
a.
wrong side

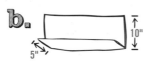
b.
10"
5"

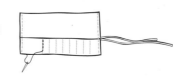
c.
18"
30"

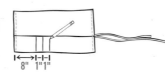
d.

e.
8" 1"1"

f.

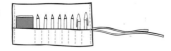
g.

56. Tank Top Tote

MATERIALS:
- tank top
- washable marker
- fabric scissors
- needle and thread
- optional decoration: soda can tabs, embroidery thread and needle

INSTRUCTIONS:
1. Turn the tank top inside out and lay it on your work surface. Draw a curved line halfway down the tank and cut along this line.

2. Sew a backstitch along this bottom edge (see diagram a).

3. Flip the tank right side out and you're finished!

4. Add soda-tab flowers to make your tank top tote look special (see diagram b).

- Decide where you'd like to have your flower on the tote, and place one soda tab at the top of that spot. Choose a bright color of embroidery thread. Sew the thread a couple of times around the middle section of the soda tab and through the front of the tote. Tie the thread in a double knot on the inside of your project.

- Then, sew one loop loosely around each corner of the soda tab and through the fabric. This is so you can adjust the position of the tab before you finish sewing it to the fabric.

- Finally, go back and sew several times around the top and bottom edges of the soda tab to add color and completely secure the tab to the tote. As long as the tab is secure, you can add as much or as little colored thread as you want. Make as many soda tab petals as you like to complete the flower. Sew a running stitch of green thread in the shape of a stem and leaves.

To make a no-sew version of this tank-top tote, cut vertical slits up the front and back from the bottom of the shirt. Pull this fringe until it curls, and then tie together the two pieces directly across from each other. Make sure the knots are tied tightly so the contents of your tote don't fall out!

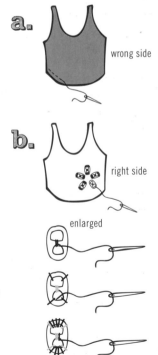

a. wrong side

b. right side

enlarged

57. Jeans Purse

MATERIALS:
- pair of jeans
- small pointy scissors or seam ripper
- fabric scissors
- needle and thread
- ruler
- washable marker or colored pencil
- embroidery thread or buttons (optional)

INSTRUCTIONS:

1. Using a pair of small scissors or a seam ripper, detach all the belt loops from jeans. Then, cut off the entire waistband below the seam and set it aside. This will be your purse strap.

2. Cut 10 inches off the bottom of one leg (see diagram a on the opposite page).

3. Turn the piece you cut in step 2 inside out and sew a backstitch in a straight line across the cut end (see diagram b). The hem of the jeans will become the top of the purse.

4. Turn the piece right side out. With your ruler, measure the width of the purse.

5. Measure and cut a panel around one of the back pockets of the jeans. This panel should be ½ inch wider than the width of your purse. Cut the bottom edge of this panel to match the shape of the pocket. Hem all sides of the cut panel except the top. To do this, fold in the edges of the pocket panel ½ inch all around (except the top). Sew a running stitch in straight lines along all the edges you folded, hemming them to the pocket panel (see diagram c).

6. Making sure the right side of the pocket panel is facing out and the bottom point is facing up, place it over the top of the purse. Sew a running stitch in a straight line along the folded edge of the pocket panel, attaching it to the top of one side of the purse (see diagram d). This panel is now the closing flap of your purse.

7. Now, it's time to attach the purse strap. Sew one end of the waistband you cut in step 2 to one side of your purse. Sew the other end of the waistband to the other side of the purse. For a super strong connection, sew the ends of the waistband to the purse with an "X" with a box around it (see diagram e).

8. You can decorate the purse if you wish. Add fun buttons, or embroider your name or some pictures on the front flap (see diagram f). Throw your new denim design over your shoulder and show off your ReMake It style.

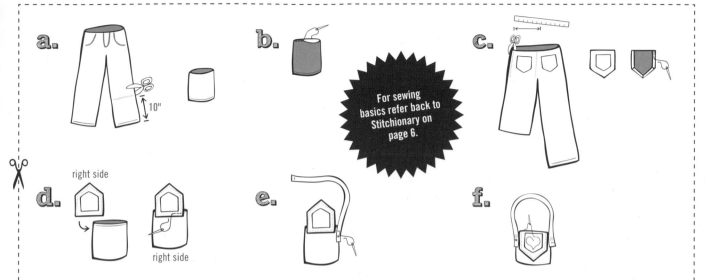

For sewing basics refer back to Stitchionary on page 6.

58. T-Shirt Skirt

MATERIALS:
- an existing skirt that fits you well (for measurements)
- measuring tape
- 3 to 5 T-shirts
- washable marker
- fabric scissors
- safety pins
- needle and thread
- shoelace, cord, or ribbon (one that is long enough to go around your waist and still tie in a knot)

INSTRUCTIONS:

1. Lay your existing skirt on a flat surface. Measure the length of the skirt down the center and add 1 inch. This extra inch will allow you to have a drawstring at the top. Then, measure the width along the bottom edge of the skirt. Take that number and multiply by 2. Divide by 5. Add 1 inch for seam allowance. Write both measurements down so you don't forget them.

2. Measure and cut 5 rectangular panels out of T-shirts to the measurements you determined in step 1. First, measure and mark the length from the

bottom of the T-shirt, leaving the hem intact. Later, this will become the hem of the skirt. Then, measure and mark the width of the rectangular panel. Cut along the lines you've marked. Before you cut, though, make sure you include the interesting part of the T-shirt on the panel.

3. Lay out the panels in the order you'd like to stitch them together. Pin each panel to its neighbor along the length of the fabric (see diagram a). Make sure the wrong side is facing out and the hems are all at the bottom. The hems along the bottom should all match up. If there is extra fabric at the top, you can always trim that later. Do this for all 5 panels to create a ring of panels with the wrong side facing out.

4. Starting at the bottom hem of each panel, sew straight lines along the edges where the panels touch (see diagram b).

5. The skirt should still have the wrong side facing out. Fold the top down by 1 inch and pin it down. Sew a straight line ¾ inch down from the top edge all around the skirt (see diagram c). This is the casing for the drawstring.

6. Flip your skirt right side out. Determine where you want the front and center of the skirt and cut a ½-inch vertical slit at the top of the skirt, creating an opening for the drawstring. Thread the shoelace through the slit and push it all the way through the drawstring casing and back out the same slit (see diagram d). Put on your ReMade skirt and tie it up!

Why not make a themed skirt? Collect shirts from summer camp or school sports teams and sew them all together!

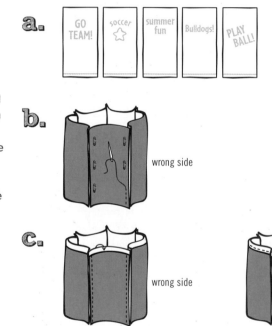

59. Cargo Pants Yoga Mat Bag

MATERIALS:
- pair of cargo pants
- fabric scissors
- ruler
- iron
- safety pins
- needle and thread
- 75 inches of cord
- cord lock

If the drawstring opening frays, put a thin line of liquid seam sealant—found at most fabric stores— on the edges of the material.

INSTRUCTIONS:

1. Start with a pair of cargo pants that is at least 28 inches long on the inseam (the inseam is the measurement from the inside top of the leg to the cuff). Cut off one pant leg at the top where the inseam begins.

2. Cut a 2-inch slit down the top of the pant leg on the side opposite the cargo pocket. This is where the drawstring will come out.

3. Flip the pant leg inside out. Fold the top edge down ½ inch, and with an **adult's supervision**, iron along the fold to make a crease. Fold the top edge down again by 1 inch, and pin it to the pant leg with safety pins (see diagram a on the next page).

4. Sew a running stitch along this fold, ¾ inch from the top (see diagram b). You now have a drawstring casing.

5. Cut a piece of cord 75 inches long. Attach a safety pin to one end, and thread it through the slit made in step 2. Thread the cord through the drawstring casing, using the safety pin to push the cord along and back out through the same hole. With both ends of the cord sticking out, insert the cord ends through the cord lock (see diagram c).

6. With the wrong side of the pant leg still facing out, cut squares 1½ inches x 1½ inches out of both bottom corners of the pant leg (see diagram d). These will become box corners, just like for the **#51 Sports Jersey Tote**.

7. Stitch along the bottom edge of the pant leg, making sure not to stitch along the squares you just cut out (see diagram e).

8. Pull the two ends of the cord through the center of the pant leg and lead them all the way down to the bottom. Thread the cord through the open box corner on the same side as the drawstring opening. Leave ½ inch of cord sticking out (see diagram f).

9. Now, you will sew up the box corners you cut in step 6. Pick up the side you sewed together in step 7 and match up the side seam and bottom seam. Flatten the pant leg this way, with the side seam laying flat on top of the bottom seam. Stitch a straight line along the box corner, perpendiclar to the side and bottom stitches. Make sure you stitch over the drawstring sticking through the bottom from step 8 (see diagram g). Repeat for the other box corner (this side will not have the drawstring sticking through).

10. Flip your bag right side out. Decorate it with embroidery thread, buttons, or any other decoration you'd like. Carry your new yoga bag to class in style. It will surprise the pants off your yogi friends!

Use the cargo pocket to carry a water bottle, keys, or other items to yoga class.

a.

wrong side

b.

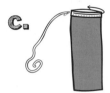

c.

d.

e.

f.

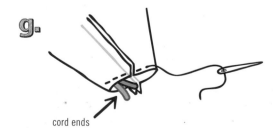

g.

cord ends

60. T-Shirt Checkerboard

MATERIALS:

- 24 bottle caps (12 each of 2 different colors)
- permanent markers and stickers (optional)
- 2 T-shirts in contrasting colors
- ruler
- fabric scissors
- fabric glue
- needle and thread
- 2 shoelaces

INSTRUCTIONS:

1. Collect and clean 12 caps each of two different colors. You can find them on soda bottles, orange juice cartons, milk cartons, and other beverages. Separate the different colors into groups. Then, decorate each group with stickers and permanent markers. These will be your checker pieces.

2. Starting at the bottom of one T-shirt, measure and cut two panels that are 12 inches wide by 13 inches tall. Cut one panel from the front and one from the back of the same shirt, leaving the hem intact. The hem will later become the casing for the drawstring.

3. Take the other T-shirt, in the contrasting color, and cut 24 squares that are 1¼ inch x 1¼ inch in size (see diagram a on the next page).

4. Lay one of the large panels on a flat surface. Arrange the small squares in a checkerboard pattern. Lay the pieces on the panel, like a checkerboard would look. Place a little bit of fabric glue on the back of each square. Glue them into place (see diagram b). Let them fully dry.

Use your crafty skills to convert your checkers into chess pieces. Use stickers or markers to decorate the lids into pawns, rooks, knights, bishops, kings, and queens.

5. Optional: Sew along the edges of the squares using a zigzag stitch (see diagram c). This will just make them more secure.

6. Place the two large fabric panels together with the wrong sides facing out so you can stitch them together. Stitch a straight line along three edges of the panels, leaving the side with the hem open. Leave the top inch, where the hem is, unstitched on both sides (see diagram d).

7. Flip your pouch right side out. Take one shoelace and thread it through one of the slits, through the entire hem, and back around through the same slit. Tie the ends into a double knot. Repeat this step with a second shoelace on the opposite side. Once you've secured the second shoelace with a double knot, pull the drawstrings away from each other and you've created your game pouch (see diagram e). Keep your game pieces inside the pouch and you're ready to take this checkers game anywhere!

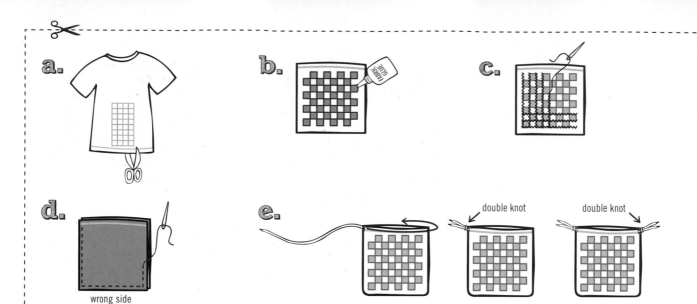

61. Comforter Laptop Cover

SKILL LEVEL:
HARD

TIME:

MATERIALS:
- comforter
- needle and thread
- 32 inches of 1-inch-wide ribbon
- scissors
- measuring tape
- safety pins

INSTRUCTIONS:

1. Determine the size of your laptop. Then, cut the appropriate-sized piece of fabric (see diagram a on the opposite page).

- 13" laptop—cut your comforter to 21 inches x 15 inches
- 15" laptop—cut your comforter to 23 inches x 17 inches
- 17" laptop—cut your comforter to 25 inches x 19 inches

Laptop size is determined by measuring the monitor diagonally. If you have a laptop size that is not indicated here, the equation to use is:

- cover width = (2 x the width of your computer) + 3 inches
- cover height = computer depth + 2 inches

A down comforter doesn't work well for this project because it can quickly turn into a big feathered mess. Use a synthetic one, or an old quilted jacket or vest works well for this project.

2. Fold in the short edges of your fabric rectangle ½ inch toward the wrong side of the fabric, and pin it down. Sew a backstitch in a straight line along these edges to make a nice hem (see diagram b).

> If you're using a sewing machine for these projects, don't use it for making your handles out of comforter fabric. It will be easier to sew by hand—some sewing machines will have trouble with so many layers at once.

3. Cut two 16-inch pieces of 1-inch-wide ribbon. Curve one of the pieces of ribbon to make a handle. Place it down on one of the short edges of the fabric, leaving 4 inches between where the two ends of ribbon handle are attached to the fabric. Pin the ribbon in place, and then sew it to the fabric. Stitch an "X" with a box around it to secure the ribbon to the fabric (see diagram c). Repeat this step with the second piece of ribbon on the other side. Make sure the two handles are directly across from each other.

Or you can use one of these alternate methods for making the handles. Use the leftovers from two neckties from project **#50 Necktie Gadget Case** or use leftover scraps from the comforter itself. To use the comforter, pin two layers of comforter together, right side out. Cut this into two 3-inch x 16-inch strips. Fold under the edges of each strip by ½ inch, and sew a straight line along each edge to hem it. Use these as your handles.

4. Fold the big piece of rectangular fabric in half with the wrong side facing out. Make sure you match up the top edges. Pin along the two open edges on the sides. Then, sew a straight line about ½ inch in along these edges (see diagram d).

5. Flip the bag right side out, and you have your very own comfy laptop cover. When you put your laptop to "sleep," you can tuck it in!

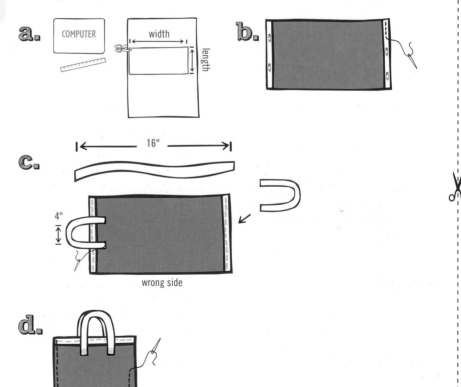

62. T-Shirt Rug

MATERIALS:

- 1 mesh sports jersey
- ruler
- 8 large T-shirts of different bright colors
- fabric scissors
- 1 large upholstery needle or 24-gauge wire

INSTRUCTIONS:

1. Cut a rectangular panel from the sports jersey that is 26 inches x 14 inches. You may need to use the front and the back of the jersey to make a piece this large.

2. Cut ¾-inch-wide strips from the large T-shirts. Your strips should be at least 20 inches long (see diagram a). You'll need about 500 strips cut from about 8 T-shirts.

3. Pull on the ends of each strip and stretch them all until their edges curl (see diagram b). Repeat steps 2 and 3 until you've cut up all of the shirts.

4. Thread one of the fabric strips through the eye of the large upholstery needle (see diagram c). It helps to poke a corner of the strip through the eye first and then pull the rest through.

> Instead of using a needle, you can bend a piece of 24-gauge wire in half and use that to pull the T-shirt strips through the rug. Be careful with this wire—ask for an **adult's** help.

5. First, you'll frame the sports jersey with fabric strips. Lay the jersey piece on a flat surface. Starting from the top corner, thread the fabric strip in through one hole and out through the hole just below it. Pull the strip through until there are 2½ inches of T-shirt strip coming out of each hole. Cut the strip so that there is an equal amount of T-shirt strip

on both sides of the hole (see diagram d). Tie the two ends of the strip into a knot to secure it to the jersey. Skip one hole along the top of the jersey, and thread the fabric through the next hole. Your needle should already be threaded with the remainder of the last T-shirt strip. By skipping holes, your rug won't be overcrowded with fabric strips. Repeat this step all the way around the edge of the jersey.

6. Once you have the rug framed, go back and fill in all of the open areas with T-shirt strips, working in horizontal rows. Skip every other row down as you did across. Use the same method as you did in step 5, threading the strip through one hole and out the other, and tying it into a knot to secure it to the sports jersey. When you start a new row, clear the way by pushing loose strips away from any nearby rows (see diagram e). This is a time consuming project, but you'll feel like a real winner once you've finished your sports jersey T-shirt rug!

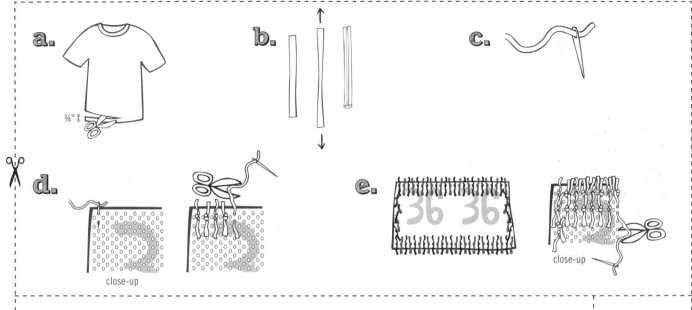

a. ¾"

b.

c.

d. close-up

e. 36 36   close-up

63. T-Shirt Scrap Flower Pin

- fabric scissors
- twist tie
- button with two holes
- pin back or safety pin

Add more T-shirt strips to make flowers thicker.

MATERIALS:
- brightly colored T-shirt
- ruler
- washable marker or pencil

INSTRUCTIONS:

1. Measure and cut ¾-inch-wide strips from the bottom of a T-shirt. From those strips, cut eight 8-inch long pieces.

2. Pull on the ends of each strip and stretch until their edges curl. Snip three holes into each strip—one in the middle and two more, each ½ inch from the ends (see diagram a on the next page). Don't cut across the whole strip. Just make small holes.

3. Slip the ends of the twist tie through a button, and pull it through so it is snug across the front of the button. Twist the ends of the twist tie together tightly until they are completely twisted at the back of the button (see diagram b).

4. Thread the end of the twist tie through the hole at the center of one of the T-shirt strips. Then, thread the twist tie through the holes on the ends of the T-shirt strip. Repeat this for all the rest of the strips (see diagram c)

5. Holding the T-shirt strips and button together, untwist the twist tie ends. Insert them through the holes in pin back or wrap them around the safety pin (whichever pin you decide to use). Adjust the twist tie until the flower and pin are held firmly in place on the pin

back (see diagram d). Remake a whole bouquet of flower pins and grow a garden on your shirt!

Mix it up and use strips from a few different shirts for a flower with different colors.

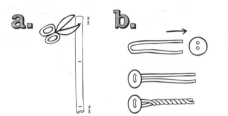

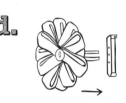

64. Scrap Fabric Circle Pin

MATERIALS:
• yogurt container
• scrap fabric (cotton or polyester works best)

• washable marker or pencil
• fabric scissors
• needle and thread
• button
• pin back or safety pin
• ruler

INSTRUCTIONS:

1. Find a yogurt container that is about 5 inches in diameter. Take the container lid and place it on top of a piece of scrap fabric. Trace around it and cut the circle out of the fabric.

2. Fold ¼ inch of the circle's cut edge in toward the wrong side of the fabric. Sew the folded fabric with a running stitch around the entire circle. Leave the ends of the thread loose (see diagram a on the opposite page).

Interesting fact: These gathered circles are known as "yo-yos," made popular in the 1800s as a way to reuse scraps of fabric.

3. Once you've stitched around the edge of the entire circle, gently pull the ends of thread until the edge of the circle gathers at the center. Tie the open ends of the thread into a double knot (see diagram b).

4. Place a button on the top of the circle and a pin back or safety pin underneath. Stitch all these pieces together so that the button, fabric circle, and pin are all connected. Pull your needle through the back of the pin, up through the fabric circle and one of the button's holes, and back down through another buttonhole, fabric, and the pin back. Do this a few times to make the pieces secure (see diagram c). Pin on your new circle accessory to keep it "a-round."

You can sew fabric circles on your clothes instead of adding the pin backs. Use these to cover up small stains.

a.

wrong side

fold

wrong side

b.

double knot

c.

button

fabric circle

pin back

65. Pocket Wall Organizer

SKILL LEVEL: MEDIUM

TIME:

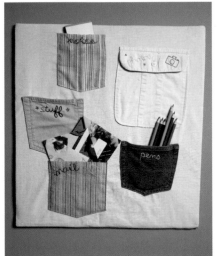

MATERIALS:
- 1 shirt with a pocket
- pockets cut out from other shirts
- measuring tape
- washable marker or colored pencil
- fabric scissors
- packaging tape
- needle and thread
- scrap cardboard at least 30 inches x 24 inches
- 6-inch piece of wire
- masking tape

INSTRUCTIONS:
1. Make sure you get permission from an **adult** before you cut up any shirts. Measure and cut the shirt into a rectangle 15 inches x 18 inches (see diagram a on the next page). You can use the back of the shirt, or you can use the front of the shirt and keep the pocket that is already there.

Save any scrap fabric pieces for another project.

2. Cut out pockets from four or five other shirts. If you want to embroider words or designs with thread on the front of the pockets, do that now (see diagram b).

3. Lay the large shirt panel you cut in step 1 onto a flat surface. Arrange the cut out pockets on top of this panel. Make sure you leave a 3-inch border around all the edges of the large panel. Pin the pockets to the large panel and sew them into place with a running stitch along bottom and side edges. Leave the pocket tops open (see diagram c).

4. Cut two pieces of scrap cardboard to 12 inches x 15 inches. Stack them on top of each other. Measure 3 inches down from the top and 5 inches in from each edge of the cardboard stack. Mark these points with a pencil. At both of these marks, poke a 6-inch piece of wire through the stack of cardboard to make a hanging hook on the loop side (see diagram d). Twist the two ends of the wire together on the opposite side. Leave the loop a little loose for hanging. Place a piece of masking tape on top of the twisted ends to keep the wire from poking

through the fabric. You'll hang your organizer from this hook when you're done.

5. Place the fabric on a flat surface with the right side facing down. Center the cardboard on top of the fabric (see diagram e).

6. Fold the edges of the fabric over the edges of the cardboard. Secure the edges in place with wide pieces of packaging tape (see diagram f). Make sure the tape holds well. Flip over your organizer, and you are ready to hang your pocket masterpiece!

a.

b.

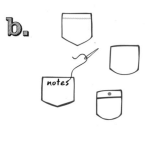

c.

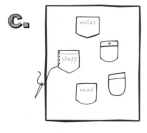

d.

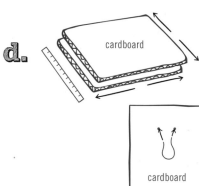

e.

f.

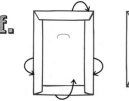

66. Felted Sweater Potholder

MATERIALS:

- 2 wool sweaters in contrasting colors
- ruler
- fabric scissors
- safety pins
- embroidery floss
- needle

INSTRUCTIONS:

1. Start by felting your two wool sweaters. To do this, ask an **adult** to wash them with hot water and a little detergent, and then dry them on high heat. Felting means shrinking your sweater on purpose. Putting wool into hot water and then into the dryer will cause it to shrink and make the fibers to bind to themselves, so they don't unravel. Felting works best with animal fibers such as wool.

2. Measure and cut a square 7 inches x 7 inches from each sweater. Cut the corners round. Then, cut one 4-inch x ¾-inch piece out of one of the sweaters. This will become the hang tab (see diagram a).

3. Stack the squares with the right sides facing out. Fold the hang tab piece in half. Insert the ends of the tab between the two squares in one corner and pin it into place. Then, pin around the edges of the stacked squares, attaching them together (see diagram b).

4. Thread your needle with embroidery floss and tie a double knot at the end. To begin the blanket stitch, which this project needs, start a stitch directly to the left of the hang tab and about ¼ inch from the edge. Refer to page 7 for instructions on doing a blanket stitch (see diagram c).

5. Continue stitching a blanket stitch around all edges of the potholder (see diagram d).

6. When you come to the hang tab, simply sew a regular running stitch through it. When you've come back to your starting point, secure both ends of the floss with a double knot (see diagram e). Show off your fancy work in the kitchen!

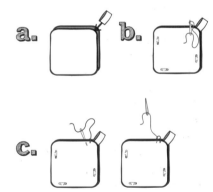

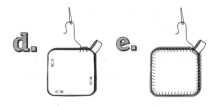

67. Felted Sweater Blanket

SKILL LEVEL:
HARD

TIME:

MATERIALS:
- 10 wool sweaters
- measuring tape
- fabric scissors
- safety pins
- needle and thread

INSTRUCTIONS:

1. Start by felting your ten wool sweaters just like you did for project **#66 Felted Sweater Potholder**.

2. Cut a 10-inch x 10-inch square from the front and back of each sweater. Then, cut 4-inch x 12-inch rectangles from the fronts and backs of all of the sleeves.

3. Lay out the sweater squares on a large flat surface and arrange them as you'd like to sew them together (see diagram a).

> Another idea: To make a no-sew quilt you can use sweatshirts instead of felted sweaters and cut a fringe around each square. Tie the pieces of the fringe together.

4. Stack squares A and B together, and pin one edge together. Sew a straight line with a running stitch along this edge of these two squares, ½ inch from the edge of the fabric (see diagram b). Flatten these two squares and lay them on your work surface.

5. Take square C, stack it on top of square B, and pin them together. Sew a running stitch along the edge directly opposite from where squares A and B are stitched together (see diagram c). This stitch should also be ½ inch from the edge. Then, stack square D, on top of square C, pin them together, and sew a running stitch along the edge opposite from where squares B and C are stitched together, ½ inch in from the edge of the fabric. Repeat steps 5 and 6 for panel groups E, F, G, H; I, J, K, L; M, N, O, P; and Q, R, S, T. When done, you'll have five sets of four panels sewn together.

6. Take rows A, B, C, D and E, F, G, H, and stack them on top of each other with the wrong sides (where the stitches show) facing out. Pin the panels together along the long edge at the bottom. Sew a running stitch along the pinned edge (see diagram d). Unpin the fabric and open it back up to lay it flat on your work surface. Repeat this step: stitch row E, F, G, H to row I, J, K, L; stitch row I, J, K, L to M, N, O, P; and finally stitch row M, N, O, P to Q, R, S, T. At the end of this step, all panels should be stitched together in one big piece.

7. To create the border, sew the sleeve pieces from step 3 into long strips. You can repeat the same steps you followed to stitch the square

pieces together, but this time, just stitch the short ends of the sleeve pieces together. Make sure the seams are all on the same side. Create two long strips that are each 4 inches wide x 37 inches long. Then, create two strips that are each 4 inches wide x 46 inches long. Cut the top and bottom of each strip in a diagonal line at a 45-degree angle (see diagram e).

8. Place the long strips down next to the blanket (see diagram f). Flip the strips over and line them up to the edges of the blanket, with the wrong side facing up. Stitch a straight line along the long edge. Repeat for other side.

9. Place the top piece along the width of the blanket with the diagonal sides together in the correct way. Stitch into place. Repeat for bottom (see diagram g).

10. Sew a straight line along the edges of the diagonal corner pieces, making sure the wrong sides are facing outward, and stitching along the back of the blanket. Make sure you do this for all four corners (see diagram h). Cozy up in your crafty new creation!

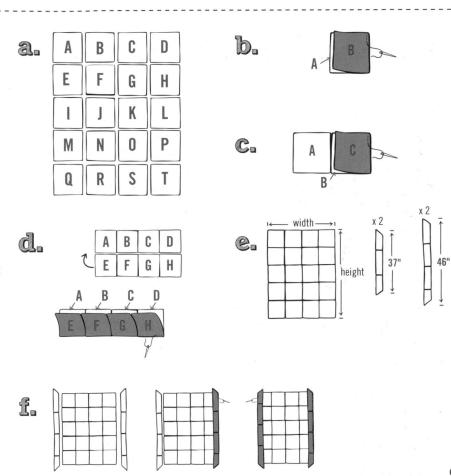

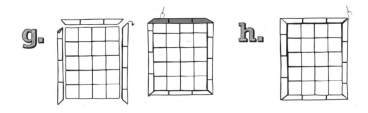

68. Felted Sweater Shrug

MATERIALS:
- wool sweater
- ruler
- chalk
- fabric scissors
- buttons (optional)
- needle (optional)
- embroidery floss (optional)

INSTRUCTIONS:

1. Choose a sweater that is too large on you. Then, felt this sweater like you did in project **#66 Felted Sweater Potholder**.

2. Lay out the sweater on a flat work surface. Cut open the sweater vertically up the center of the front.

3. Cut off the bottom of each sleeve about 5 inches from the cuff (see diagram a). Or, try on the felted sweater, and measure a couple of inches below the elbow, and mark the line with chalk. Take the sweater off and cut there.

4. Cut off about 6 inches from the bottom of the sweater (see diagram a). Or mark approximately where the sweater hits the middle of your rib cage when you try on the felted sweater. Then, take off the sweater and cut along that line. Round the edges of the front corners of the sweater.

5. Stitch around the edges with embroidery floss (this is optional). Because your sweater is felted, it won't fray, but stitching can add a nice finished look and some color (see diagram b). A blanket or running stitch looks nice.

6. Decorate the shrug with buttons or patches, as you like. You can sew a button to the front of each side of the shrug opening and use a ponytail holder to make a cute closure (see diagram c). Slip into your shrug to keep your shoulders warm.

a.

b.

c.

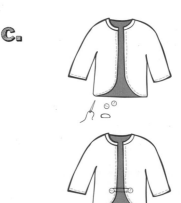

69. Felted Sweater Fingerless Gloves

MATERIALS:

- wool sweater
- ruler
- fabric scissors
- buttons (optional)
- needle (optional)
- embroidery floss (optional)

INSTRUCTIONS:

1. Start by felting your wool sweater like you did in project **#66 Felted Sweater Potholder**.

2. Cut off both sleeves of the sweater about 12 inches from the cuff (see diagram a).

3. About 1½ inches below the cuff, cut a small hole on one side. Do this for both sleeves (see diagram b).

4. Slip on the fingerless gloves. Your thumb goes through the small hole. This will help stretch them out to the right shape before you decorate them or do any decorative stitching.

5. Optional: Stitch around any cut edges. Because the sweater is felted, it shouldn't fray, but stitching will add a nice finished look and some color. A blanket or running stitch looks nice.

6. Decorate your fingerless gloves by sewing on buttons or embroidering simple shapes onto them (see diagram c). Now, you are ready for winter in a stylish, recycled way!

a.

b.

c.

70. Felted Sweater Sleeve Scarf

SKILL LEVEL:
MEDIUM

TIME: 🕐

MATERIALS:
- wool sweater
- ruler
- fabric scissors
- safety pin
- yarn
- large needle

INSTRUCTIONS:

1. Start by felting your wool sweater just like you did for project **#66 Felted Sweater Potholder**.

2. Lay out your sweater on a flat work surface. Cut from the armpit of one sleeve up to the neck. The whole sleeve should be left in one piece.

3. Measure 4 inches down from the top of the piece you just cut and mark a line with safety pins (see diagram a).

4. Sew a running stitch along this line going through both layers of the sleeve. Then, stitch a border along the edges of the sleeve through both layers and close up the bottom cuff. Cut apart the two layers up to the stitching (see diagram b). You'll have two flaps at the top of the sleeve that are the same size. You can now remove the safety pins and put them aside.

5. Stitch another line through the top open end of the sleeve. There should be a 4-inch space just below this line (see diagram c). This space will become your hole for inserting the end of the scarf.

6. Sew a running stitch all around the edges of this gap, through one layer only. This stitch is just decorative (see diagram d).

7. Optional: Turn on your craftiness to personalize your scarf. Decorate it with yarn, buttons, and fabric scraps. Tuck the end of the scarf through the gap at the top. Bundle up in your cozy new creation!

a.

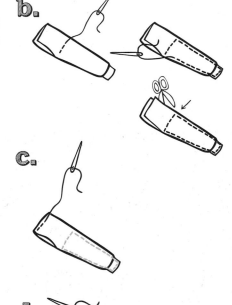

b.

c.

d.

✂

Chapter 4
Metal Makeover

71. Mint Tin Jewelry Box

MATERIALS:
- mint tin
- washable marker
- scrap piece of wood
- hammer
- awl or nail
- sandpaper
- 24- or 28-gauge wire
- scissors
- paperboard
- permanent markers
- rhinestones
- stickers

INSTRUCTIONS:

1. With your washable marker, mark on the inside of the lid of your tin where you want to punch holes. It's best to have two rows of six holes that are ½ inch apart (see diagram a on the opposite page). Then, make two marks, one on either end of the lid. Later, you will string wire through these holes to hang dangly earrings.

2. Rest the lid facedown on top of a scrap piece of wood to protect your work surface. Place your awl or nail on one of the points you marked and hammer down. Be careful when you do this, and ask for an **adult's supervision**. Repeat this step for all the points you marked on the lid (see diagram b). You can smooth out any rough edges with a bit of sandpaper.

3. Thread each end of a piece of wire through the two holes at either end of the lid. Twist the ends of the wire together on the top side of the lid (see diagram c). Cut off any extra wire with strong scissors or wire cutters. Make sure you have an **adult's supervision** when you do this. Cover the twisted piece of wire with a piece of clear tape.

4. Cut two dividers from pieces of paperboard, like a tissue box, to place into the bottom of the tin. Measure the paperboard to fit the width of your tin and cut them as needed. Fold and crease the paperboard lengthwise, then insert them into the bottom of your mint tin (see diagram d).

5. Decorate the outside of your tin with permanent markers, rhinestones, stickers, and any other decorative items you like (see diagram e). Place post earrings through the holes in the lid, and string dangly earrings onto the wire. Place rings and necklaces in the compartments on the bottom. This jewelry box project is an open-and-shut case!

 **a.**

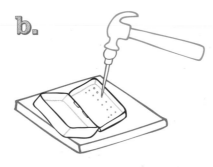 **b.**

 twist

 c.

 d.

 **e.** EMMA

Decorate the outside of your tin with permanent markers and stickers.

72. Tin Tic-Tac-Toe

MATERIALS:
- magazines
- glue stick
- flexible magnet (like the kind local stores use for advertising or found in craft stores)
- mint tin
- scissors
- ruler

INSTRUCTIONS:

1. Cut "X" and "O" letters from magazine pages or create your own letters (see diagram a on the next page). You'll need five of each.

2. Glue the letters onto the flexible magnet (see diagram b). Use your scissors to carefully cut apart the game pieces (see diagram c).

3. Now, it's time to create the game board. Cut two strips from a magazine that are 2 inches x ¼ inch. Then, cut two pieces of magnet the same size. First lay the two magazine strips vertically on the inside of the mint tin lid. Then, lay down the two magnet strips horizontally over those to keep

them in place (see diagram d). Stick the game pieces to your mint tin, grab a friend, and start playing tin tic-tac-toe.

a.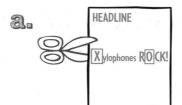

b.

c.

d.

73. Mint Tin Word Play

Make themed word-play games using eco-friendly, sports, or fashion words. Then decorate the outside of the tin to match the theme.

MATERIALS:
- magazines
- flexible magnet (like the kind local stores use for advertising or found in craft stores)
- scissors
- glue stick
- mint tin

INSTRUCTIONS:
1. Cut out words from magazine pages until you have enough to cover the entire surface of the flexible magnet.

2. Use your glue stick to attach all the magazine words onto the magnet.

3. Carefully cut apart the pieces. Then, stick the pieces to your mint tin and rearrange the words to create sentences and fun phrases. Send a fun message to a friend this way!

74. Fruit Cup Candle Holder

MATERIALS:

- metal fruit cup
- pliers
- masking tape
- washable marker
- towel
- awl or nail
- hammer
- ¹⁄₁₆-inch paper hole punch (optional)
- permanent markers

INSTRUCTIONS:

1. Remove any labels from the outside of the metal cup. Use pliers to press down any frayed or jagged edges along the inner top edge of the cup (see diagram a).

2. Cover the outside of the metal cup with masking tape wherever you want to make a dot design. Draw a fun dotted design on the masking tape with your washable marker (see diagram b).

3. Fill your cup with water and place it in the freezer until the water is fully frozen (see diagram c). This prevents the metal from denting while you punch holes. It also gives you a harder surface to work on.

4. Place your cup on its side on top of a towel. This will help keep it from slipping and will soak up the water as the ice starts to melt. Have an **adult** hold the cup in place while you hammer the awl or nail into the places where you've marked the dot design (see diagram d).

5. Remove the masking tape and decorate the outside of your cup using permanent markers (see diagram e). Place a candle in the metal cup, and let it shine!

Always be careful with candles. Never light a candle without an **adult's supervision** or leave one burning when you're not around.

a.

If you have a small ¹⁄₁₆-inch paper hole punch, use that to punch the holes to avoid steps 3 and 4.

b.

c.

d.

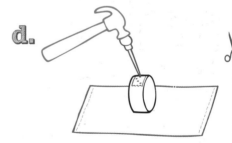

e.

75. Soda Can Tab Belt

MATERIALS:

- 110 soda can tabs, the rough edges cut off or sanded away
- 100-inch-long shoelace or nylon cord
- scissors
- measuring tape

INSTRUCTIONS:

1. With **adult supervision**, bend down, cut off with wire cutters, or sand down any rough edges on the soda tabs.

2. Tie a loop at the center of the cord that is about 1 inch in diameter. Secure it with a tight double knot. The entire loop should be about the same size as a soda tab. Later, this loop will become the belt closure.

3. Thread the one end of the cord through one of the soda tab holes, and thread the other end of the cord through the other soda tab hole. Always thread from back to front of the tabs (see diagram a).

4. Place a second tab on top of the first. Bring the ends of the cord over one side and through the middle of this second tab. Then, bring the ends of the cord back through the middle of the first tab (see diagram b).

Measure: The belt in this project is for someone with a waist size of 25 inches to 30 inches. You can scale it up or down if you need a different size. For every 4 inches of belt, 13 tabs are needed.

5. Put a third tab under the second tab. Then, bring the ends of the cord through one side of the third tab and then up through the center of the second tab. You'll be stacking your soda tabs like two layers of bricks. The top layer will overlap the bottom layer by a half tab (see diagram c).

6. Repeat steps 4 and 5 until the belt is a few inches longer than your waist size (see diagram d).

7. Tie the end of the cord in a tight double knot and trim the loose ends (see diagram e).

8. To wear the belt, just slip the knotted end through the loop you tied at the beginning (see diagram f). If you're thirsty for more soda tab projects, try the same technique to make a bracelet.

 **a.**

 **b.**
1 2   1 2

c.
1 2   3   1 2 3

d.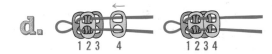
1 2 3   4   1 2 3 4

e. ←double knot

f.

76. Paint Can Planter

MATERIALS:

- empty paint can
- paint brush
- paint
- rocks
- potted plant that's smaller than 5½ inches in diameter

INSTRUCTIONS:

1. If there's paint left inside the can, use that paint to paint the outside of the can. Let any additional paint in the can harden and dry. Remove any labels from the outside of the can and paint it. Save the lid for project **#77 Paint Can Lid Clock**.

2. Fill the paint can halfway with rocks.

3. Place your potted plant inside the can. Fill up the rest of the can with rocks around the sides of the potted plant. You'll want the top of the plant pot to line up with the top of the paint can. Now, rock out with your new, low-cost planter!

77. Paint Can Lid Clock

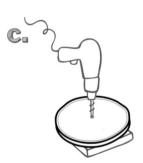
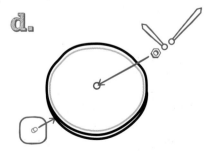

2. With the painted side of the lid facing up, find the exact center of the lid and mark this spot with a marker (see diagram b).

3. With the painted side still facing up, place a scrap piece of wood under the lid. Have an **adult** drill a hole through the center mark, using a ⅜-inch drill bit (see diagram c).

4. Now, attach the clock mechanism to the lid. Insert the clock mechanism from the back of the lid through the hole in the paint can, screw the mechanism into place, and then add the minute and second hands and the batteries (see diagram d). Have an **adult** help you hang your new clock on the wall if you have hanging hardware, or lean it against the wall on your dresser. Now, it's time to sit back and look at your finished project.

MATERIALS:
- paint can lid
- paint brush
- permanent marker
- scrap piece of wood
- drill and ⅜-inch drill bit
- clock mechanism and batteries

Have an *adult* help you with this project.

INSTRUCTIONS:
1. Paint the bottom of a paint can lid and let it fully dry (see diagram a).

Design idea: Use a marker or magazine page cut outs to mark numbers on the clock face, like in project **#45 Record Clock.**

78. Metal Lid Photo Frame

MATERIALS:

- 3 metal lids
- marker
- scrap piece of wood
- hammer
- awl or nail
- scissors
- photos
- 28 inches of ⅛-inch-wide ribbon
- ruler
- glue stick
- glitter glue

INSTRUCTIONS:

1. On the inside rim of one lid, mark two dots 1 inch apart. Then, mark two more dots 1 inch apart on the exact opposite side of the inside rim. Repeat this step on another lid, again marking four dots. On the third lid, mark only two dots 1 inch apart (see diagram a on the next page).

2. Place the lid on its edge against the scrap piece of wood. Place the awl or nail against one of the dots. With an **adult's supervision**, hammer down through the lid. Repeat this step until all of the dots are punched through (see diagram b).

3. Make sure you have permission to cut up a photo, or use a photocopy of a photo. Place one lid on top of one of your photos, and trace around the edge. Remove the lid and draw another circle inside the first circle that is about ¼ inch smaller. Cut along the smaller line. Test to see if the photo fits inside the lid (see diagram c). If it doesn't quite fit, trim around the edges until it does. It's OK if the edges aren't perfect—you will decorate them with glitter glue. Repeat step 3 with the other two lids. When you do this step, pay attention to where the holes are and make sure when you hang your lids, your photos will be straight.

5. Glue the photos inside the lids. Cover the edges of the photos with a border of glitter glue, and allow them to fully dry (see diagram d).

6. Line up your lids vertically in the order you wish to hang them. The lid with only two holes goes on the bottom. Cut and string 12 inches of ribbon through top two holes of the top lid. Tie a bow at the top (see diagram e).

7. Cut and string an 8-inch piece of ribbon through the top two holes of the second lid. Then, string the ends of the ribbon through the bottom two holes of the top lid. Adjust the ribbon until the lids are about 1 inch apart from each other. Then, tie the ribbon into a double knot and make a bow. Repeat this step to attach the second and third lids together (see diagram f). Trim off the extra ribbon and hang your masterpiece from the top bow.

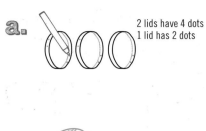

a. 2 lids have 4 dots
1 lid has 2 dots

b.

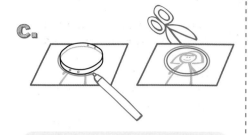

c.

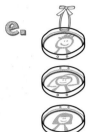

d.

GLUE STICK

GLITTER GLUE

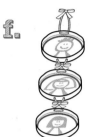

e.

f.

This makes a great holiday gift! Surprise your relatives or friends with photos of you and some of their favorite people. Or you can use a group of hanging photo frames to make a photo family tree.

79. Tin Can Desk Organizer

SKILL LEVEL:
MEDIUM
TIME:

MATERIALS:
- 3 tin cans
- 2 plastic bottles
- marker
- pushpin
- scissors
- a piece of flat, flexible plastic

- 15-inch piece of elastic cord
- tape
- decorative paper (optional)

INSTRUCTIONS:

1. Clean and wash three empty tin cans. Find two plastic bottles that fit snugly inside the cans. Clean those, too.

2. Place a bottle inside one of the tin cans. Mark a line where the bottle comes out of the can.

3. Use the pushpin to punch a hole right above the line. Slip your scissors into the hole, and carefully cut along your marked line.

4. Mark a vertical line along the length of the bottle base right down the middle, around the bottom, and back up the other side. Cut along the line. You will now have exactly two halves of a bottle base, lengthwise.

5. Place the open end of the bottle base on top of the flat plastic, and trace around the rim with your marker. Draw a

straight line across the plastic from one edge of the curved line to the other (see diagram a). Cut along the half circle you traced, and tape the piece of plastic to the open end of the bottle base (see diagram b).

6. Repeat steps 2 through 5 with the other plastic bottle and two cans.

7. Fold the elastic cord in half, and tie it around one of the cans. Secure the cord with a double

knot. Then, take the open ends of the cord and tie them around the next can, again securing with a double knot. Tie the ends of the cord around the third can and double knot them (see diagram c).

8. Decorate the cans and bottles with decorative paper, glitter glue, and any other art supplies you have. Turn the stack of cans on its side, with two of the cans on the bottom, one on the top. Insert

the plastic bottle drawers (see diagram d), and you've made yourself an organizer!

a.

b.

c.

d.

80. Umbrella Frame Hanger

MATERIALS:
- umbrella frame
- wire cutters
- masking tape (optional)
- 24- or 28-gauge wire
- ribbon (optional)
- several "S" hooks (optional)

Have an *adult* supervise you for this project.

INSTRUCTIONS:

1. If your umbrella frame still has fabric on it, remove it (see diagram a on the next page).

2. Open the umbrella frame, and place it upside down on your work surface.

3. Working all the way around, cut away the bottom tension wire in each umbrella spoke with wire cutters (see diagram b). Be careful working with the wire. Don't worry if there are dangling ends—you can tuck these back into the umbrella frame, securing them with a piece of tape. Make sure an **adult** is there to help.

4. With the tension wires removed, you'll be able to move the main spokes freely. Find the small hole in the end of one spoke. Cut a 12-inch piece of wire, and loop it through this hole. Fasten the wire securely to the hole by twisting or wrapping it (see diagram c).

5. Working with every other spoke, thread a 12-inch piece of wire through each end hole. Fold these spokes up toward the center of the umbrella.

They should rest against the center pole and each other. Cross each spoke on top of its neighbor, only crossing in one direction. Wrap each wire tightly to the center pole to secure the spokes (see diagram d).

6. Repeat step 5 with the other umbrella spokes, crossing each one on top of its neighbor in the same direction and wrapping the wires around the center pole (see diagram e).

7. Trim any extra wire. If the umbrella has a curved handle, you have a ready made hook to hang it. If not, you can use ribbon to make a hanging loop from the umbrella handle, and hang it from the ceiling or wall. Hang your scarves, jewelry, and more on your new attractive, sturdy frame! This is a great rainy-day project.

Add a few "S" hooks so that you can hang purses or hats from the frame.

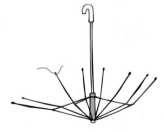

a.

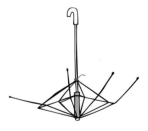

b.

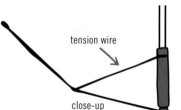

tension wire

close-up

c.

d.

e.

81. Bottle Cap Magnets

MATERIALS:

- metal bottle caps
- tweezers
- ruler
- scrap paper (take-out menus, magazines, newspapers, or sheet music)
- super glue
- small stickers
- glitter glue pen
- clear nail polish
- strong magnet that's about ¾ inches in diameter

INSTRUCTIONS:

1. Use tweezers to pull out the center rubber piece on the underside of each bottle cap (see diagram a).

2. Cut out a circle with a 1-inch diameter from a piece of scrap paper. With an **adult's supervision**, super glue this onto the underside of the cap and let it dry (see diagram b).

3. Decorate this glued piece of paper with other bits of paper and stickers. Use the glitter glue pen to squeeze some glitter around the edge (see diagram c). Let it fully dry.

4. Brush a coat of clear nail polish on top of the paper and let it dry (see diagram d).

5. With an **adult's supervision**, attach the magnet to the back of the cap with super glue, and let it set for 24 hours before you use it (see diagram e). Make as many as you want. Then, put your new magnets on display—don't keep them bottled up!

a.

b.

c.

d.

e.

82. Bottle Cap Necklace

MATERIALS:
- bottle cap
- scrap piece of wood
- awl or nail
- hammer
- tweezers
- jump ring
- clear nail polish
- glitter glue
- scrap paper (take-out menus, magazines, newspapers, or sheet music)
- super glue
- necklace

INSTRUCTIONS:

1. Place a bottle cap sideways on a scrap piece of wood. With an **adult's supervision**, place your awl or nail against the inside rim of the cap, and carefully hammer down, punching a hole into it (see diagram a). If you have a ¹⁄₁₆-inch paper hole punch, you can use it to punch this hole.

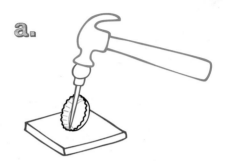

2. Use tweezers to open a jump ring, and thread it through the hole you just punched. Close the jump ring with the tweezers (see diagram b).

3. Follow steps 1 through 4 from project **#81 Bottle Cap Magnets** to decorate the inside of your bottle cap.

4. Thread a necklace or lanyard through the jump ring (see diagram c). Your friends will flip their lids when they find out you've made this super-cool necklace.

Chapter 5
Glorious Glass

83. Eyeglass Picture Frames

SKILL LEVEL: EASY

TIME: 🕐

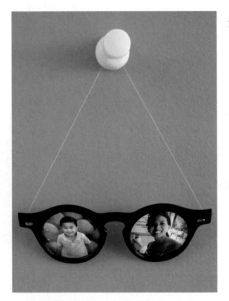

MATERIALS:
- old pair of eyeglasses
- mini screwdriver
- blank scrap paper
- pencil
- scissors
- photos
- clear tape
- 15 inches of fishing line

INSTRUCTIONS:

1. Remove the arms of the eyeglasses with the mini screwdriver (see diagram a). A mini screwdriver can be found in an eyeglass repair kit.

2. Make a template to cut out your photos. On a piece of scrap paper, trace around one lens of the eyeglasses. Cut along the line you drew (see diagram b). Hold the template behind the glass lens to see if it fits properly. If not, trim it until it does.

3. Use the template to cut out your photos. Trace around the template onto your photo, and cut it out (see diagram c). Because eyeglass lenses are the same shape and mirror images of each other, you don't have to make a whole new template for the other lens. Just flip the template over, and use it to trace around and cut a photo to fit in the other eyeglass lens.

4. Use a small amount of clear tape to tape your photos to the back of the eyeglasses (see diagram d).

5. To make a hanging loop, thread the fishing line through each side of the glasses where the arms were once attached, and tie the ends into knots (see diagram e). Hang your eyeglass picture frame on your wall, and focus on your new creation!

a.

b.

c.

d.

e.

84. Baby Food Jar Candle Holder

INSTRUCTIONS:

1. Measure and cut a rectangle out of scrap paper that is about 7¼ inches x 1¼ inches. Create your own design onto the rectangle and cut out any shapes you choose (see diagram a). The shapes can be flowers, stars, circles, or anything you like.

2. Attach the piece of paper to the outside of the jar with glue or clear tape (see diagram b). Add your candle and shine some light onto your new project.

a.

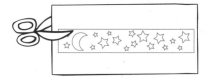

b.

MATERIALS:
- scrap paper (take-out menus, magazines, newspapers, or sheet music)
- ruler
- scissors
- baby-food jars, cleaned and labels removed
- glue or clear tape

Always be careful with candles. Never light a candle without an **adult's supervision** or leave one burning when you're not around.

85. Baby Food Jar Organizer

MATERIALS:

- 6 small baby-food jars, cleaned and labels removed
- marker
- cardboard
- hammer
- nail or awl
- pushpin
- used tennis ball
- 6 screws (1 inch or shorter; preferably wood screws)
- screwdriver
- paint or decoupage materials (optional)

To make a fun, rolling organizer, just use two small jars.

INSTRUCTIONS:

1. Mark two holes in the center of one baby-food jar lid, about 1 inch apart. Cut small pieces of scrap cardboard to place under the lids while you make the holes. The cardboard will protect your work surface. With an **adult's supervision**, use the hammer and nail to carefully punch holes where you've made the marks (see diagram a on the opposite page).

2. With a pushpin, poke two matching starter holes into the top of your tennis ball. Then, insert a screw into the tennis ball and screw it firmly into place to enlarge the hole (see diagram b). You can use your lid as a guide to place these holes.

3. Repeat steps 1 and 2 with the other five baby-food jars. For a guide on how to place the holes, see diagram c. Use the seams on the ball for reference. Think of it this way: if the tennis ball were a cube, you'd put two holes through each side. Remove all the screws from the tennis ball.

4. Before you do the next step, you can decorate the outside of the lids with paint, stickers, or colorful paper (see diagram d). Just make sure you don't cover up the holes.

5. Line up one jar lid with one set of holes in the tennis ball. Attach the lid (with the top of the lid facing the tennis ball) firmly with two screws. Repeat this step with the other five lids (see diagram e).

7. Fill each jar with whatever you want, and screw them into their lids. You'll really have a ball with this organizer!

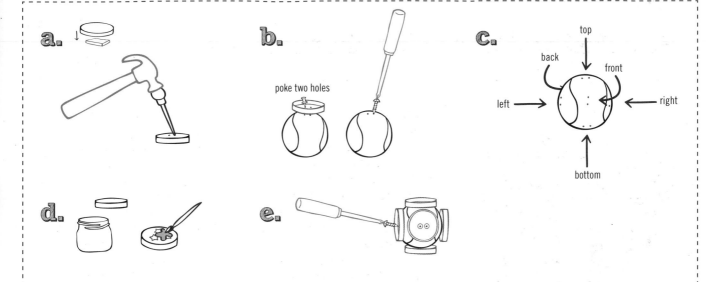

86. Glass Bottle Vase

MATERIALS:
- 40 inches of 24-gauge wire
- glass bottle
- masking tape
- wire cutters
- about 50 beads
- round-nosed pliers

INSTRUCTIONS:
1. Wrap one end of the wire around the rim of the bottle three times. Then, twist the short end around the longer end a few times to keep it from unraveling (see diagram a on the next page). Tape the wire to the bottle to keep it from shifting while you're working with the beads.

2. String ten beads onto the wire. Push the beads to 3 inches from the top of the bottle. With an **adult's supervision**, grip the wire with your round-nosed pliers just below the beads. Using the tip of the pliers make a small, tight loop with the wire to keep the beads from slipping down (see diagram b).

3. Continue beading this way, leaving 3 inches between the top bead and the loop, and then placing a loop below the strand of ten beads. Do this until you've reached the end of the wire.

Use a combination of colors, sizes, and shapes of beads to decorate your vase.

4. Wrap the beaded wire around your bottle. Secure the end of the wire by wrapping it around another part of the wire a few times (see diagram c). Remove the piece of tape from the top of the bottle. Then, fill the vase with water, insert a flower, and give it to a friend to brighten his or her day!

87. Salt And Pepper Shakers

MATERIALS:
- 2 clean small jars
- masking tape
- pen or pencil
- scissors
- scrap cardboard
- awl or nail
- hammer

INSTRUCTIONS:

1. Remove the lids from the small jars. Draw the letters "S" and "P" in dots onto pieces of masking tape. Apply one to each jar lid (see diagram a on the opposite page).

2. Cut a few small pieces of scrap cardboard to place under the lids while you hammer out the letters "S" and "P." The cardboard cushions will reduce rough metal edges on the other side and protect your work surface (see diagram b).

3. With an **adult's supervision**, use your hammer and awl or nail to make holes in the lids through the tape you've marked your dots. Go slowly and gently to make sure the

holes are all the same size (see diagram c).

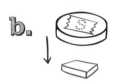

4. Remove the tape and wash and dry the lids. Fill your jars with salt and pepper, and spice up your table with your new shakers.

Use ideas from other projects to decorate the outside of your salt and pepper shakers like using paper templates in projects **#84 Baby Food Jar Candle Holder**.

88. Jelly Jar Drinking Glass

SKILL LEVEL:
HARD
TIME: 🕐🕐

MATERIALS:
- glass jar (jelly or salsa jars work great)
- electrical tape
- pen
- craft knife
- cloth
- glass-etching cream (etchall® is non-toxic and available in craft stores)
- paintbrush

Ask an adult to help you with this project.

INSTRUCTIONS:
1. Make sure your glass jar is clean and dry, and labels and glue are completely removed.

2. Make stencils from electrical tape. Draw shapes and letters on top of electrical tape with a pen. Then, ask an **adult** to help you carefully cut out your stencil with a craft knife (see diagram a on the next page). You can stencil your name, your favorite shapes, or numbers.

3. Apply your tape stencil to the jar. With a cloth between your hands and the jar (so you don't get fingerprints on it), smooth down the tape stencil so there are no gaps or air bubbles. Use long strips of tape to divide areas and make stripes on your jars (see diagram b).

4. With an **adult's supervision**, apply glass-etching cream with a clean paintbrush (see diagram c). Follow the instructions for use on the packaging. Once you've applied the etching cream, you may need to let it sit for as long as 5 minutes. So be patient, and don't touch your project.

etchall® is a non-toxic product and is safe for children to use with adult supervision. Try using it for other glass projects that you want to personalize!

5. Follow the instructions on the etching-cream package for washing the cream from the glass (see diagram d). When the glass is wet, it's hard to see the etched area. Don't worry— if you followed the instructions from the etching cream, it worked just fine!

6. Remove the tape stencil and wash the glass one more time with soap and water. Then, enjoy a beverage in your new, personalized drinking glass!

a.
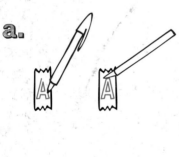

b.
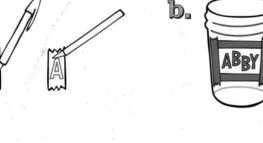

c.
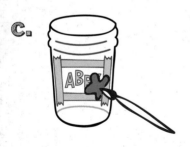

d.

Chapter 6
Junk Pile Jazz-Up

89. Crayon Caddy

MATERIALS:

- 25 crayons
- wide packing tape
- yogurt cup
- 2 pieces of 16-inch fishing line or dental floss

INSTRUCTIONS:

1. Lay the crayons down on your work surface in a line. Nudge them all into place so their bottom edges are lined up and the sides touch each other.

2. Lay a strip of tape down on top of your row of crayons to fasten them together.

Stand up the row, and make a circle with the crayons. The tape should be on the inside. Overlap the tape inside to secure the circle tightly (see diagram a). Now, you have a "tube" of crayons.

3. Turn the yogurt cup facedown on the table. Slip the crayon tube over the cup, with the crayon bottoms facing up (see diagram b).

4. Slip the middle of the fishing line over the bottom of one of the crayons about ½ inch down from the crayon bottom. Weave the fishing line back and forth around the crayons crisscrossing both ends. You will eventually come back to the original starting spot. Tie a tight double knot with the ends of the string and trim off the extra (see diagram c on the opposite page). Then, tape down any loose edges to the inside of the tube.

5. Slip the yogurt cup out of the circle. Flip the crayon circle over and repeat step 4 about ½ inch down from the top

of the shortest crayon (see diagram d).

6. Cut off the bottom of the yogurt cup about 2 inches from the bottom. Save the top part for another project like **#30 Plastic Butter Tub Buttons**, and slip the bottom of the yogurt cup into the crayon circle. The yogurt bottom will keep the crayons in shape and become the bottom of the pencil holder (see diagram e). Fill your new arty organizer with desk supplies, paintbrushes, or anything else you like!

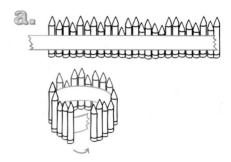

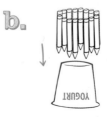

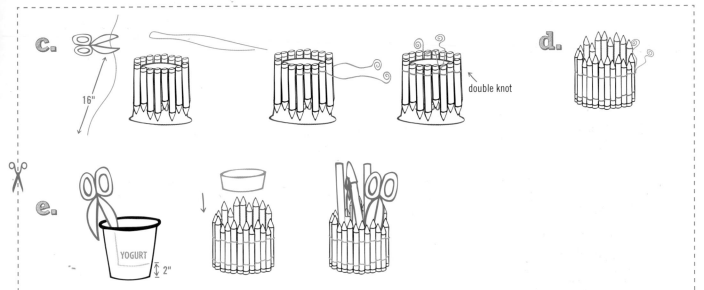

c. 16"

double knot

d.

e.

YOGURT

2"

90. Puzzle Friendship Necklace

SKILL LEVEL: MEDIUM
TIME:

MATERIALS:

- 2 connecting puzzle pieces
- pushpin
- 2 large jump rings
- tweezers
- 2 necklaces or 2 20-inch pieces of lanyard
- clear fingernail polish

INSTRUCTIONS:

1. Poke a hole through the top part of each puzzle piece. Wriggle the pin to make the hole bigger (see diagram a).

2. Use tweezers to open up one jump ring. Thread the jump ring through the hole you created with the pushpin. Close up the jump ring. Repeat this step with the other jump ring and puzzle piece (see diagram b).

3. Thread one necklace or lanyard through each jump ring.

Carefully coat both sides of each puzzle piece with a thin layer of clear nail polish. Hang the necklaces and let the polish dry. This ReMade version of the split-heart best-friends necklace will help you and your best friend stay connected!

a. b.

91. Chopstick Jewelry Organizer

MATERIALS:

- 6-ounce yogurt cup
- scissors
- 6 chopsticks
- rubber band
- ball of yarn
- masking tape
- white glue

INSTRUCTIONS:

1. Cut the top lip off the yogurt cup. Trim any rough edges off the lip with your scissors, and then set it back down onto the body of the cup. You'll use the body of the cup for working on this project, but it won't be part of the final organizer.

2. Loosely group five of the chopsticks together with a rubber band about 2 inches away from the pointy ends. Put them into the yogurt cup with the pointy ends down. Then, twist out the chopsticks into a spiral. The lip of the yogurt cup will form a ring around the group of chopsticks.

3. Push the sixth chopstick, pointy end down, into the center of the group to keep the five chopsticks from falling down. Arrange the other five chopsticks so they are evenly spaced and resting against the loose lip of the yogurt cup. All of the pointy ends should be touching the bottom of the cup, and the center chopstick should be touching the center of the cup's bottom (see diagram a on the opposite page). The sixth chopstick is only to help keep the rest in place. It will be taken out later on.

4. Tie the yarn onto one of the five chopsticks and the yogurt cup lip in an "X" shape. Loosely tape the rest of the chopsticks to the yogurt cup lip (see diagram b).

5. Fasten the chopstick to the ring by wrapping the yarn around it and the yogurt cup lip a few more times in an "X." Then, wrap the yarn firmly around the yogurt cup ring until you reach the next chopstick. Remove the tape from that chopstick, wrap the yarn around that chopstick and the yogurt cup lip in another tight "X" a few times. Repeat this step moving along the entire yogurt cup ring (see diagram c). If you come to the end of the piece of yarn, tie it into a tight knot with a new piece of yarn. Trim the loose ends of the yarn, and tuck them under the wrapping as you go.

6. After wrapping the fifth (and last) chopstick and completely covering the yogurt cup lip with yarn, tie a tight knot around the ring and cut off the extra yarn (see diagram d).

7. Remove the sixth chopstick from the center of the bundle and remove the yogurt cup bottom from the jewelry organizer. Tie yarn next to the rubber band near the bottom of the chopsticks. Once the

yarn is tight, take away the rubber band.

8. Make sure everything is even and set, and then reinforce with white glue all around the chopsticks where they touch the cup ring. Glue down any stray ends of yarn. Let the glue dry completely. Flip the organizer so the wider side is facing down (see diagram e). Place rings, bracelets, and other pretty things onto your new jewelry organizer.

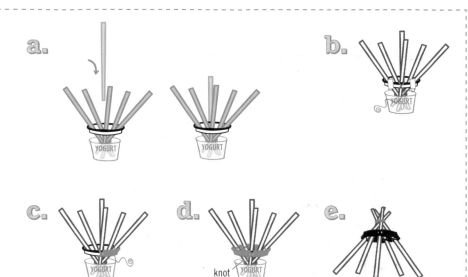

92. Dishware Cake Stand

SKILL LEVEL:
EASY

TIME: 🕐

MATERIALS:
- plate
- dessert cup with a stem
- washable marker
- silicone or clear caulk (ask an **adult** for some you can use)

INSTRUCTIONS:

1. Place the plate upside-down on a flat surface. Center the dessert cup upside-down on top of the plate. Trace around the outside of the dessert cup with the washable marker.

2. Remove the dessert cup. Squeeze a line of silicone or caulk along the inside of the circle you marked.

3. Place the dessert cup back in the center of the plate right onto the silicone or caulk. Let it dry for 24 hours. Wasn't this ReMake It project a piece of cake?

It's best to hand wash your new cake plate so you don't risk it breaking in the dishwasher.

93. Junk Pile Picture Frame

MATERIALS:

- pile of junk
- picture frame
- hot glue gun
- masking tape and scrap paper
- or plastic
- newspaper
- paint primer
- paint
- sponge brush or paintbrush

INSTRUCTIONS:

1. Collect a pile of junk to decorate your frame (see diagram a). This is a great way to use up lots of odds and ends and turn them into a cool piece of artwork!

2. Ask for an **adult's help** when using a hot glue gun. Glue your pile of junk to the picture frame (see diagram b). Lay out all of your pieces before you glue them to see just how you want the finished frame to look. You can glue your junk in one layer or go for a real 3-D effect by piling items on top of each other. Just make sure each piece is securely glued on.

> Some ideas for your pile of junk: old coins, keys, used pens, caps, pencil stubs, mixture of hardware, old or broken jewelry, single earrings, crumpled foil and paper, plastic toys, soda pull tabs, buttons, cut-up credit cards

3. Remove the glass from the frame. Place a piece of scrap paper inside the frame to protect the backing. Place the project on a surface covered with newspaper. Cover the decorated frame with a couple coats of paint primer.

4. Finally, paint the whole thing in one color with a sponge brush or paint brush. Let the paint dry fully. Then, take off the protective paper, and replace the glass. You'll be amazed at the way this project changes your trash into a beautiful treasure—worthy of framing!

a.

b.

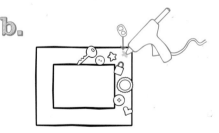

94. Book Recycling Bin

MATERIALS:

- 2 hardcover books that are the same size (picture books are great for this project)
- craft knife (ask for an **adult's** help!)
- clear packing tape
- large piece of corrugated cardboard
- marker, pen, or pencil

INSTRUCTIONS:

1. Strip out the pages of the books (and save these for scrap paper projects!). Most books are usually attached at a couple of spots that you can slice through with a craft knife (have an **adult** help you with this). If you turn the book on its end, it's easy to see where you need to cut to remove the book's pages.

2. Stack the books covers, one on top of the other, with their insides facing in. Tape the short edges of the books together with one long piece of tape down the edges (see diagram a).

3. Now, set the taped-together books upright on your work surface, and arrange them into a diamond shape. Reinforce where the book edges meet with more tape on the inside (see diagram b).

4. Now, make the bottom of your trashcan. Set the books on top of a piece of scrap cardboard. Trace on the cardboard along the bottom edge of the books (see diagram c).

5. Cut out a cardboard base along the traced lines. Slip the books over the cardboard base. It should fit snugly. Tape along all edges of the base where it meets the books, both inside and out (see diagram d).

Why settle for an ordinary wastebasket when you can have one with a story?

a.

b.

c.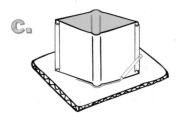

d.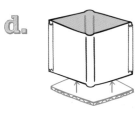

125

95. Book Purse

MATERIALS:

- old hardcover book
- craft knife (ask for an **adult's help!**)
- ruler
- pencil
- scrap paper
- scrap fabric
- scissors
- iron
- fabric glue
- wide ribbon
- binder clips
- masking tape

INSTRUCTIONS:

1. Strip out the pages of the book (and save these for scrap paper projects!). Most books are usually attached at a couple of spots that you can slice through with a craft knife (have an **adult** help you with this). If you turn the book on its end, you can see where you need to cut to remove the pages.

2. Stand the book up on its end and measure the length and width of the top. Measure and draw a vertical rectangle this size onto a piece of scrap paper. Turn the paper so the bottom of the rectangle is near you. Then, draw a triangle on each side of the rectangle by adding a line 3 inches long on each side of the top of the rectangle and drawing diagonal lines to the bottom corners of the rectangle. Then, draw a frame around the shape you just made by adding 1½ inches around the entire outside of it (see diagram a on the opposite page).

3. Using this pattern, trace and cut two pieces of fabric for the purse side panels. Fold in all of the edges of the fabric by ½ inch. Glue these down to the wrong side of the fabric to make a hem (see diagram b).

4. Fold both side pieces in half with hemmed sides facing out.

With an **adult's supervision**, run a hot iron over the fold to make a crease (see diagram c).

5. Open the book cover and lay it on a flat surface with the inside facing up. Lay both panels of fabric on the flat surface, one above and one below of the book cover. Overlap one side of the fabric panel and the top edge of the book cover by 1 inch, making sure the top corner of the fabric meets the top corner of the book. Glue the fabric onto the book cover. Repeat this step with the bottom fabric panel and the bottom edge of the book cover (see diagram d).

6. Allow the glue to dry, and then apply two horizontal 1-inch stripes of glue to the other side of the book cover, one along the top edge and one along the bottom edge (see diagram e).

7. Close the book cover and press the other edges of the fabric panels to the other edges of the cover. Use the end of a pencil to smooth the fabric onto the inside of the cover.

Insert something between the two sides of the book cover to keep the panels separated while the glue dries.

8. Cut a piece of ribbon for your strap. For a 5-inch x 7-inch book, you'd need a 45-inch-long ribbon. You should cut yours longer or shorter depending on the size of your book.

9. Stand the book on its spine. On the outside of the book cover, measure 1½ inches in from the right edge. Mark this spot and use your ruler and pencil to draw a vertical line around the front, spine, and back of the cover (see diagram

f). Then, put glue along this pencil line on your book cover. (Stand the book back on one end so the glue won't smear on your work surface.)

10. Fold the ribbon in half to find its center. Lay the center point of the ribbon on the center point of the spine of the book over the line of glue. Pull the ribbon tight and clip it together with binder clips at the top (see diagram g). Let the glue dry completely.

11. Draw another vertical line 1½ inches in from the other end of the cover. Stand the cover on the other end, and put glue along this pencil line. Unclip

the two ends of the ribbon. Bend one ribbon in curve, and glue it down to the glue line on the other end of the book cover and halfway across the spine (see diagram h). Hold it down with tape if necessary. Let glue the dry completely.

12. Now, curve and glue the other ribbon to the glue lines on the other side of the book cover (see diagram i). The two ends of the ribbon should meet at the spine of the book. Make sure there is enough glue on the spine to secure the ends of the ribbon. Tape this end down while it dries. Show off your knowledge with this smart book purse!

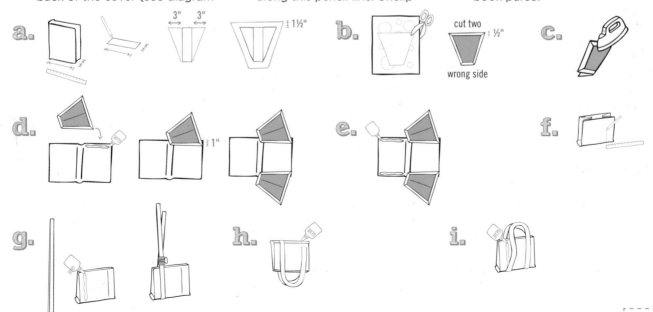

Author Bio

TIFFANY THREADGOULD is a design junkie who gives materials a second life. As a kid she turned gum wrappers into bracelets (just like project #23), empty jars into vases for her mom's flowers (just like project #86), and more. All grown up and still ReMaking It, Tiffany runs a business called RePlayGround (www.replayground.com). There you can find ReMake It recycling kits, learn about more DIY project ideas, or even contact her to chat about trashy crafts!

Acknowledgments

An overflowing scrap pile of gratitude goes out to everyone who helped ReMake this book. To my editor, Alli Shaloum Brydon for seeing the potential in what others consider scrap. And many thanks to the entire Sterling crew and photographer Kevin Schaefer, too!

An enormous heartfelt thanks to Abby Kelly who played a huge design role in the ReMaking of this book. A big trashy thank you also goes out to Carly Miller, to my superstar family, everyone at TerraCycle, and to my dear friends who have been my constant support and have seen me through garbage thick and thin. Thank you all!